Mare
Oost
ZEE Svevicum
Bornholm
Rugen
D
Pruyssen
Pomeren
Polonia
Bohemia
Austria
Bavaria
Croatia
Slavonia
Dalmatia
Hungaria
Transsylva
Moldavia
Pontus
Graecis M. Nigr.
Propontidem
Lituania
Podolia
Sarmatia
Bulgaria
Thracia
Macedonia
Morea
Pelopo nelus
Mare Tyrrhenum et Inferum
Golfo di Venetia
olim Superum Adriaticum et mare
ITALIA
Roma
Sicilia
Salerno
Archi
Riga
Smolensko
Kioff
Cracow
Leopolis
Wien
Salſburg
Venetia
Padua
Ravenna
Florenta
Ancona
Spoleto
Ostia
Capua
Taranto
Brindisi
Otranto
Constantinop
Adrianopoli
Nicopolis
Sophia
Philippopoli
Larissa
Theſsalonica
Athenae
Corintho
Napoli
Cefalonia
Zante
Corfu
Valona
Durazzo
Ragusi
Scutari
Belgradum
Buda
Presburg
Temeswar
Hermenſtat
Cibinium
Danubius
Berlin
Dantzik
Elbing
Stetin
Roſtock
Lubeck
Hamburg
Lunenburg
Braunſwyck
Brandenburg
Leypzig
Breſlau
Olmutz
Nuerenburch
Regenſpurg
Ingolſtat
Lauingen
Auſpurg
Memmingen
Insprugk
Trent
Aquileia
Istria
Zara
Sibenico
Spalato
Warſow
Minsko
Lublin
Chelm
Lutzko
Belz
Bar
Orihau
Kilia
Deſter
Meſembria
Roma
S(Pers)a
Samos
Andri

Ivory Diptych Sundials 1570–1750

To Dan and Sue Chipchase –
Soli Deo Gloria!

Steve Lloyd
August 1992

The Catalogue of the Collection of Historical
Scientific Instruments, Harvard University

William J. H. Andrewes, Project Director
Bruce Chandler, General Editor

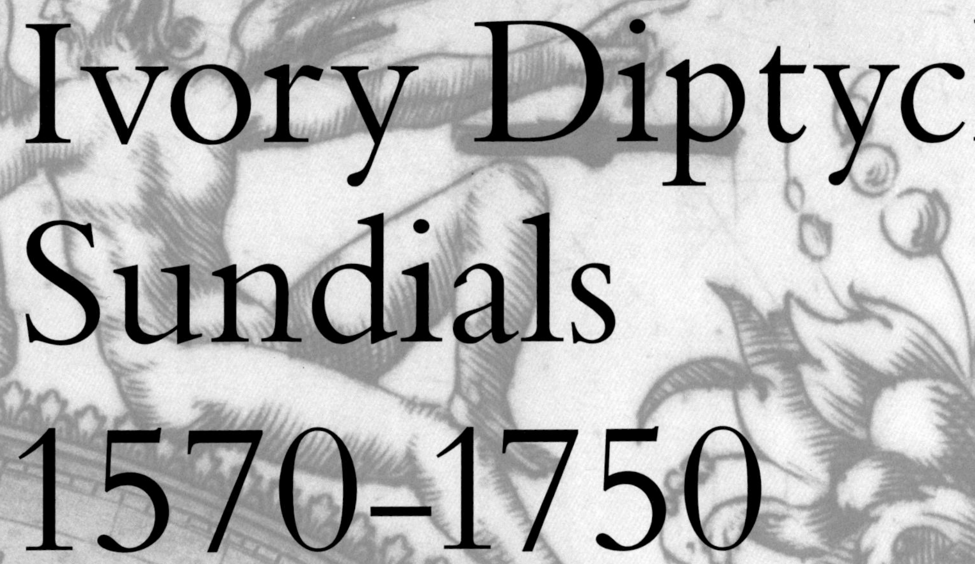

Ivory Diptych Sundials 1570–1750

Steven A. Lloyd

with Introductions by
Penelope Gouk
and A. J. Turner

The Collection of Historical Scientific Instruments,
Harvard University

Distributed by
Harvard University Press
Cambridge, Massachusetts, and London, England
1992

The Collection of Historical Scientific
Instruments, Harvard University

Advisory Committee:
William J. H. Andrewes (David P. Wheatland
Curator), Clark A. Elliott, Owen J. Gingerich,
David S. Landes, Abdelhamid I. Sabra,
Sidney Verba, David P. Wheatland,
Edward O. Wilson, Christoph Wolff

The Catalogue of the Collection of
Historical Scientific Instruments
Ivory Diptych Sundials, 1570–1750

Photography: Hillel S. Burger
Design: Pamela Geismar/Sisman Design
Printing: Toppan Printing Company
Distribution: Harvard University Press
Publisher: Collection of Historical
  Scientific Instruments

Printed on a pH-neutral paper.

Library of Congress Cataloguing-in-Publication Data

Harvard University. Collection of Historical
Scientific Instuments.
    Ivory Diptych Sundials, 1570–1750/ Steven A. Lloyd
  with introductions by Penelope Gouk and A. J. Turner.
      p. cm. — (The Collection of Historical Scientific
  Instruments, Harvard University) ·
    "Distributed for the Collection of Historical Scientific
  Instruments, Harvard University."
    Includes bibliographical references and index.
    ISBN 0-674-46977-1
    1. Sundials—Exhibitions.   2. Sundials—Catalogues.
  3. Diptychs—Exhibitions. I. Lloyd, Steven A.
  II. Title. III. Series: Collection of Historical Scientific
  Instruments, Harvard University (Series)
  QB215.H37 1992   91–29703
  681.1'11—dc20   CIP

Contents

Foreword

"My goodness! You never told me they looked like *that!*" The reaction of the Director of the Yale Art Gallery to the small ivory sundials from the Harvard Collection on exhibition in his museum in 1964 is not unusual for someone who has never seen these attractive instruments before. To anyone not acquainted with the many different kinds of instruments designed to tell the time by the sun, the word *sundial* may conjure up an image of the common garden sundial. The Director no doubt had entered the exhibition with the depressing thought of seeing a large number of these in a row.

It is surprising that scientific instruments are appreciated by so few people, because they provide such a wealth of information. They tell us what information was important to the people who used them, what materials were available and fashionable at the time, and what sorts of limitations were presented by the science and technology of the day. And each piece displays the ingenuity and skill of its maker. In short, scientific instruments are a remarkable resource for the historian. There is a good reason behind the design and construction of every part of a scientific instrument. For example, why were the sundials presented in this catalogue made of ivory? The answer, which may at first appear to be based purely on decorative grounds, becomes apparent when the instrument is used. In addition to improving the visibility of the gnomon's shadow by day, the whiteness of the ivory allows the time to be determined at night by the light of the moon.

The man primarily responsible for assembling this extensive collection of sundials is David P. Wheatland, Harvard Class of 1922, whose story I will relate shortly. Of the 82 ivory diptych sundials described and illustrated in this catalogue, 76 come from Wheatland's private collection: 31 of these were part of the Joseph Drecker collection in Leiden, Holland, which Wheatland acquired in 1959 with the assistance of R. Newton Mayall. Five of the instruments in this catalogue come from the collection of Dr. Harold C. Ernst of Boston, one of the first professors of bacteriology in the United States. (Ernst also assembled a large collection of microscopes, which was displayed for many years in Building D of the Harvard Medical School and is now on permanent loan to the Collection of Historical Scientific Instruments.) Ernst's collection of approximately 140 portable sundials, most of them Chinese or Japanese and covering a period from about 1600 to 1921, was displayed at the Boston Museum of

Fine Arts before it was given to Harvard University in 1938. For many years these sundials were exhibited in the Transparency Room at the Harvard College Observatory. Twelve instruments from this collection were stolen from the Observatory Library on June 11, 1979; however, nine were mysteriously returned to the Observatory mail room on March 24, 1980. One of the three instruments not returned was a small ivory and wood diptych dial, described as being French with a calendar on the outer surface. The entire Ernst collection was transferred to the Collection of Historical Scientific Instruments on June 16, 1980. The magnificent magnetic azimuth sundial made by Charles Bloud in 1653 (cat. no. 56) was donated by Mrs. Irving T. Snyder of San Diego, California, in 1974, with the encouragement of Agnes Mongan, the Curator of Drawings at the Fogg Art Museum.

The ivory dials in this collection have been exhibited on several occasions in the past. Twenty-one were included in the March 1964 exhibition at the Yale University Art Gallery, which was organized in collaboration with Derek de Solla Price. In October 1964 and in April 1967, similar exhibitions were mounted in the Widener Room at the Widener Library at Harvard. From September 24 to November 12, 1990, an exhibition of all but two of the instruments in this catalogue was held at the Houghton Library at Harvard. The opening of this exhibition coincided with the Tenth International Scientific Instrument Commission Symposium.

As already mentioned, David Wheatland, to whom we owe the existence of this collection, was graduated from Harvard in 1922. He entered his family's Maine forestry business, but his interests lay more in physics. After about six years he returned to Harvard to work for Professor E. Leon Chaffee as a research associate at the Cruft Laboratory. Toward the end of World War II, he became the first operator of the mammoth computer, the IBM Mark I. His interest in collecting early scientific books, especially those relating to electricity and magnetism, was stimulated by a visit to England about 1937. As a result, he decided to collect books on this subject printed before the year 1800. Having thought his target would be approximately 100 volumes, he was somewhat discouraged when he was shown a catalogue listing of 1,000 titles in this field. Within ten years he had acquired about 3,000 books on the subject, but he knew by this time that his eventual goal was closer to 5,000.

After the war ended, Harvard began discarding a large quantity of scientific apparatus from its laboratories. Wheatland thought that some of the more significant pieces should be preserved, because one day they would serve as a useful historical resource. In a short time, his office was filled with apparatus, and it became apparent that other arrangements had to be made. As a result of a dinner meeting on May 6, 1947, attended by Samuel Eliot Morison, Paul H. Buck, E. Leon Chaffee, Reginald Fitz, Theodore Lyman, William A. Jackson, I. B. Cohen, and David P. Wheatland, recommendations were made to appoint a committee to organize a public exhibition of the instruments and to bring in proposals for their future care. David Wheatland, Samuel Eliot Morison, and I. B. Cohen were subsequently appointed to this committee, with Wheatland as Chairman. The first exhibition of the instruments, which was originally planned to be held in November of that year to coincide with the centenary of the Lawrence Scientific School, was postponed until February 1949, when a public exhibition was opened in the foyer of the Mallinckrodt Chemical Laboratory. At this time, the name "The Collection of Historical Scientific Instruments" was adopted, and Wheatland was appointed Curator, a title he retained until May 1964, when he became Honorary Curator. Through the offices of Provost Paul Buck, space was provided in the Semitic Museum basement, and a sum of $1,000 was granted to cover the costs of the exhibition and to develop the Collection. Because Harvard had no funds available to maintain a study collection of this type, Mr. Wheatland was obliged to establish an endowment from his personal resources and from donations from friends to finance most of the expenses. At present, the day-to-day operation of the Collection is supported entirely from this endowment.

It was soon after becoming Curator that David Wheatland began his personal collection of scientific instruments. In this effort, he was strongly encouraged and assisted by his wife, Elizabeth H. Wheatland. They formed his collection with the intention that one day his instruments would be used to fill in gaps in the Harvard Collection and thereby provide a more complete resource for teaching and research. In 1949 he acquired a large number of sundials and other instruments from a private collector in Germantown, Pennsylvania, 23 of which are included in this catalogue. Twelve more ivory diptych sundials were purchased in France in 1952. In 1959 he purchased the Drecker collection, which included 243 sundials and about 350 books. Wheatland's collecting career, which continued over the next thirty years, was spurred by a fascination with all types of scientific instruments and with books relating to their history, construction, and use. His collection of over 700 sundials and about 700 books on dialing, both dating from the sixteenth to the twentieth century, would, by itself, have been a remarkable achievement for any individual. But his interest extended beyond these to instruments used for astronomy, navigation, surveying, calculating, timekeeping, and many other disciplines. It was in the history of electricity, however, that his collection of instruments and books was most concentrated. His system of tagging each piece immediately with a label describing details of its acquisition and origin, which he developed right from the start of his collecting career, has become a kind of hallmark for the objects he acquired.

In the early 1980s, Wheatland began to make some decisions on the future of his personal collection. He had already given a large number of books to Harvard in 1967 but, although the Houghton Library had established a policy for acquiring such rare items, the Collection of Historical Scientific Instruments had severe limitations of space and had retained its original policy of collecting only those instruments that had been used at the University. Because of Harvard's lack of interest at the time, Wheatland invited his friends Roderick and Marjorie Webster to select any instruments they wanted for the remarkable collection of astronomical instruments they were building at the Adler Planetarium in Chicago. About the same time, he sent a large number of instruments to auction at Christie's, New York. He retained the majority of his collection, though, and donated some of the instruments to Harvard in 1985. In 1988 those instruments not sold at auction were given to the Collection of Historical Scientific Instruments, for by this time a new policy to develop the Collection for teaching and research had been adopted by the University.

This catalogue is the fourth publication on the Collection of Historical Scientific Instruments and the first in a series of volumes describing specific types of instruments in detail. The first book about the Collection, *A Catalogue of Some Early Scientific Instruments at Harvard University,* was issued in 1949 in connection with the exhibition at the Mallinckrodt Chemical Laboratory. One year later, I. B. Cohen published an important account of the early history of science at Harvard entitled *Some Early Tools of American Science.* This included, in Appendix III, the earlier catalogue describing 46 pieces from the eighteenth and early nineteenth centuries. In 1969, David Wheatland produced a valuable reference work entitled *The Apparatus of Science at Harvard, 1765–1800.* This work described and illustrated 84 instruments and included information on their purchase and use. Because these three publications were all concerned with the early apparatus used at Harvard, some people have gained the impression that the Collection contains only instruments used at Harvard during the late eighteenth and early nineteenth centuries. In fact, during the last forty years, the Collection has developed one of the most remarkable and diverse holdings of scientific instruments in the world. Currently the inventory lists approximately 15,000 instruments dating from about 1500 to the present. A library with over 9,000 titles, containing books and pamphlets relating to the instruments, has also been assembled.

The present series of catalogues is being prepared with two objectives in mind: the first is to describe briefly the operation and the history of the instruments included; the second is to provide detailed descriptions and illustrations of each piece for the purpose of research and comparison with similar artifacts from other collections. For this reason, the catalogue is divided into sections, with the artifacts separated into appropriate groups according to their country of origin and type. A technical section describes the operation of these sundials, and introductory essays place each group of instruments in their historical context. The photographs allow the reader to study each instrument in detail.

The costs of producing a well-designed and heavily illustrated book are high, but the intrinsic value of this type of work would be diminished if those who would find it useful are unable to afford it. Therefore, the production cost of each catalogue has to be subsidized. To find funding for this type of work is not easy, and I am indebted to John B. Robinson and his father, John R. Robinson, for their encouragement, support, and generosity. Through the kind intervention of Dean Phyllis Keller, Harvard University has also contributed to this publication.

Museum catalogues become a labor of love, and many have given their time and energy to this project. Steven Lloyd, who expressed his interest in the sundials soon after my arrival at Harvard in September 1987, has, in addition to writing this catalogue, devoted many hours to developing the Double Helix database in which information on each object is organized. Sherry Stacy of Apple Computer, Inc., came to our rescue three years ago by lending us a Mac Plus computer; although slow by today's standards, it has nonetheless enabled us to arrange and sort the inventory in a manner that would otherwise have been impossible. Bruce Chandler, who was General Editor of the catalogues of The Time Museum, Rockford, Illinois, has strongly supported this project. Lucy Sisman, Pamela Geismar, Hillel S. Burger, Alan Gilpatrick, and Kate Schmit have devoted many hours above and beyond their original expectations of the work this book would involve.

Penelope Gouk has served as our main advisor on the text and has written the historical introduction to the Nuremberg dials. Anthony Turner kindly agreed, at very short notice, to contribute the historical introductions to the sections on the dials made in France and Flanders. Roderick and Marjorie Webster, Clare Vincent, and Owen Gingerich have given generously of their time in reviewing and commenting on the manuscript before publication. Thanks are also due to Clark Elliott, who helped to establish the history of the Ernst collection; to Ebenezer Gay, my predecessor, who has been a valuable source of information on several occasions; to William Newman, who made the original translation of the booklet describing the use of the magnetic azimuth dials; to Danielle Hanrahan and Peter Schilling, who identified the woods used in the construction of some of the dials; and to Howard Boyer of Harvard University Press for his advice on details of production and distribution. Among the many others who have helped directly or indirectly with this project, I am indebted to the Advisory Committee of the Collection of Historical Scientific Instruments, whose guidance has been of inestimable value. Last, and perhaps foremost, I would like to thank David Wheatland, whose kindness and generosity will continue to enrich the lives of many for generations to come.

William J. H. Andrewes
David P. Wheatland Curator
21 December 1991

Preface

This catalogue differs from most other catalogues of sundial collections in a significant way — photographs of each side, as well as a composite view, are included for each object. The number of pictures may seem extravagant, but we think it is important to publish a complete photographic catalogue of at least one major collection of ivory diptych sundials that is both large and diverse enough to provide an adequate basis for comparison with the objects in other collections. Because the Wheatland Collection provides an adequate cross section of the various styles of ivory diptychs, as well as several unique specimens, a complete photographic record of these specimens is particularly worthwhile. The "extravagance" in photography is not without limits, however; budgetary constraints necessitate the use of duotone rather than color photographs.

The recent publication of two catalogues consisting in part of ivory diptych sundials (by the Whipple Museum of the History of Science, Cambridge, England, and the Kunstgewerbesammlung/Stiftung Huelsmann, Bielefeld) and Penelope Gouk's book, *The Ivory Sundials of Nuremberg,* indicates that there is considerable current interest in the topic. Although brief introductory and historical essays are included here to help orient the reader, the format is that of a catalogue rather than a scholarly treatise. As such, I would heartily encourage the interested reader to refer to scholarly texts on the subject for further details; for the sake of brevity I have summarized material extensively covered elsewhere and added detail only where needed to supplement current literature on the subject.

Diptych sundials are hand-held portable timepieces made from two thin tablets (either of ivory, bone, wood, brass or other metal, or some combination thereof) hinged together at one end. Diptych dials (*diptych* means "folded in two") are typically rectangular, hinged at the narrow end, although square, oval, and octagonal diptychs are not uncommon. A few diptychs were manufactured to resemble tiny books and are therefore hinged lengthwise.

Diptych sundials contain small clasps to fasten the dial shut for carrying, and often additional clasps to hold the dial open so that the two tablets are fixed perpendicular to each other when in use. With a few rare exceptions, most diptych dials have a string gnomon (shadow-caster), which stretches at an angle (determined by the latitude of the observer) between the two tablets. When held level and oriented north-south with the aid of the compass in the bottom tablet, the shadow of the string gnomon indicates the local apparent solar time on one or more hour scales on the lower (horizontal) tablet. If the position of the string is permanently fixed (in other words, not adjustable by the user), one commonly finds an additional (redundant) hour scale on the vertical tablet.

Most diptych sundials are better thought of as astronomical compendia. In addition to indicating the local apparent solar time, many diptychs have features that allow one to determine: 1. the time at night, either by using a lunar volvelle or moondial to convert the "local apparent lunar time" as cast by the shadow of moonlight to the corresponding local apparent solar time, or by using a nocturnal to indicate the time from the observed position of the Big or Little Dipper relative to the pole star, Polaris; 2. the time of the high and low tides, by means of a simplified lunar volvelle known as a tidal volvelle; 3. the length of the day or, equivalently, the number of hours of daylight and darkness during each month; 4. the current astrological sign or month; 5. the altitude of celestial or terrestrial objects, by the use of a quadrant; 6. magnetic north, which during the mid to late seventeenth century conveniently lined up closely with true geographic north for large portions of central and western Europe; 7. wind direction, used to indicate prevailing wind and weather patterns for a given fixed location; 8. the latitudes of various important cities throughout Europe and the Near East; and 9. the yearly epacts corresponding to the Julian and/or Gregorian calendars, which aid in the calculation of the dates for the celebration of Easter and other moveable feasts according to the Roman liturgical calendar. The reader may consult the relevant portions of Section 1 or the glossary for definitions of the dial furniture and for explanations of their correct usage.

Each of the 82 sundials catalogued here is shown in a perspective photograph, the direction of which was determined by the need to show various clasps or attachments along the edges of the tablets, as well as in four head-on views, one shot of each face of the diptych. The system of labeling the faces of the tablets is that used by Zinner (1956; reprint, 1979) and subsequently adopted by recent authors and auction catalogues. Face *Ia* refers to the outer face of the upper tablet, *Ib* to the inner face of the upper tablet, *IIa* to the inner face of the lower tablet (which, with few exceptions, includes the main compass bowl), and *IIb* to the outer face of the lower tablet. For the majority of the dials, these faces are arranged with the hinged sides of each face in the center, so that they are seen in their respective positions when the dial is open. However, in certain cases, for legibility, the images have been inverted.

Rectangular, oval, and octagonal diptychs are hinged along the narrower edge unless otherwise noted. Metal clasps are referred to by their distinctive shapes; they may be horse-head (see, for example, cat. no. 16), scythe (cat. no. 55), or simple wire hook clasps (cat. no. 27). Auxiliary clasps are found on several dials to enclose pin gnomon rods, wind vanes, or other accessories in hollowed-out recesses in the bottom tablet. Diptychs with auxiliary equinoctial and/or polar dials (predominantly the French diptychs) have a hinged metal arm, usually recessed in the right side of the lower tablet, for adjusting the angle of the upper tablet to the latitude or co-latitude. The labeling on the compass bowls of the

Nuremberg diptychs is engraved directly into the ivory; in contrast, French, Italian, and other diptychs usually have a very thin sliver of ivory at the base of the compass bowl, making engraving difficult, if not impossible. While the Nuremberg *Kompassmachers* were prohibited from using any sort of paper inserts, the French and other non-German diptychs typically have a printed or hand-labeled paper or metal insert in the base of the compass bowl, marking the cardinal directions and occasionally additional information. Where indicated in the compass bowl, magnetic declination (the difference in degrees between geographic and magnetic north) is given to the nearest degree. The original glass or mica covering the compass bowl is often missing or has been replaced; on several dials it is difficult to determine whether the glass is original. A metal ring was normally used to fix the glass compass cover in place. Most of the extant compass needles appear to be original; obvious replacements have been noted in the descriptions of the dials. Most other auxiliary devices have been lost, with the exception of a Nuremberg wind vane (cat. no. 34, see figure 6, p. 29) and two French pin gnomon rods (cat. nos. 63 and 76, see figure 1, p. 24). Incomplete dials (especially several Bloud-type diptychs) have been included for the sake of completeness and to allow for comparison of decorative motifs.

The dials have been assigned makers and dates by the following criteria, in this order: 1. inscriptions of makers' names and/or dates; 2. inscriptions of initials, characteristic makers' marks, or distinctive decorative motifs; 3. comparison with similar dials in our collection and others worldwide; and 4. internal evidence of the years of intended use, such as epact tables. Comparison was made with diptychs in the following museums: Smithsonian Institution, Washington; Adler Planetarium, Chicago; The Time Museum, Rockford, Illinois; Museum of the History of Science, Oxford; Whipple Museum of the History of Science, Cambridge; British Museum, London; Science Museum, London; National Maritime Museum, Greenwich; Kunstgewerbesammlung, Bielefeld; Astronomisch-Physikalisches Kabinett, Kassel; Germanisches Museum, Nuremberg; and Bayerisches Nationalmuseum, Munich. I am indebted to Penelope Gouk for providing access to her private photographic files and notes on many of the Nuremberg diptychs in various European museums. Although the instruments consulted for the sake of comparison do not represent all sixteenth- and seventeenth-century diptych dials, they do account for the majority of ivory diptychs accessible to the public. A checklist of ivory diptychs in public collections is currently being compiled by the author.

Where the dials were signed, the description gives the name in the same style (upper/lower case) and spelling as found on the instrument. A word of caution should be mentioned with respect to the Nuremberg dials. The letters *d* and *t* are interchangeable in the *Mittelfränkisch* dialect spoken around Nuremberg, and I have adopted the earlier spelling *Ducher* in preference to *Tucher* for this family of craftsmen, since both spellings were used at various dates. Variations in spelling are, at best, an ambiguous means of distinguishing one craftsman from his kinsman.

The dimensions given are nominal overall dimensions, because the warping of the ivory in many cases made estimates of the original dimensions difficult. The dimensions are those of the ivory alone and do not include brass clasps, hinges, or other protruding metal pieces. Many of the German dials have brass "bun" feet, which are simply round spacers often placed on the corners of the lower tablet that allow the dial to rest on a level surface without scratching the ivory or metal volvelle.

Several diptychs have old string gnomons, but few of these appear to be original. Where it was practical to do so, string gnomons were inserted when photographing the perspective shots and removed for clarity when photographing the inner faces of the tablets. Likewise, the glass compass bowl covers were removed, when it was feasible (leaving the metal ring in place), to avoid unnecessary reflections during photography. Occasionally the string gnomon or glass disc was so tightly held in place that no attempt was made to remove or replace it.

Although all of the photographs in this catalogue are in duotone, references are made to colored engravings and the condition of the ivory. Little color other than red and black was used before the 1610s. The reader is referred to Penelope Gouk's *Ivory Sundials of Nuremberg* for several excellent color photographs of many highly decorated Nuremberg diptychs. Many of the early French dials have painted scenes or colored backgrounds on face Ib, often augmented with gilt floral designs. The dark background of these French diptychs makes them often very difficult to read; on some the paint has begun to peel and chip off. Later Bloud-type French diptychs are highly decorated, but only black coloring was used on the ivory. The paper compass bowl inserts were often painted bright colors.

The cataloguing of Harvard's collection of ivory diptych sundials on a computerized database began in the context of a course I was taking toward a Master's degree in the History of Science Department in the fall of 1987. While the emphasis of this project has changed over the past three years, its two main goals have remained constant: to develop a computerized database of the ivory sundials as a prototype for cataloguing the entire collection, and to facilitate the publication of catalogues of the collection's holdings both in a standard printed format and eventually also on computer disc.

I wish to extend many thanks to a former Sunday school teacher, Sandy Rutiser, who first aroused my interest in the art, science, and craft of dialing.

I am particularly grateful to Hillel S. Burger for his patience, endurance, and skill throughout the past year of photography. His talent for lighting has rendered several of the photographs considerably more legible than the original object as seen in standard room lighting. I would also like to thank the curator, Will Andrewes, for his encouragement and support over the past three years. Additionally, I am indebted to Alison Sandman for her encouragement and help in preparing the dials for photography, and to Sherilyn Hausey of the Boston Latin Academy for assistance in translating various Latin inscriptions.

I am also grateful to those who graciously and patiently showed me their collections of ivory diptychs at the following museums: The Time Museum, Rockford, Illinois; the Adler Planetarium, Chicago,

Illinois; the Smithsonian Institution, Washington, D. C.; the Whipple Museum of the History of Science, Cambridge, England; the Science Museum, London; and the Astronomisch-Physikalisches Kabinett, Kassel. In particular, I wish to thank: Karon Anderson, Jim Bennett, Sarah Genuth, Peggy Kidwell, Uta Merzbach, Carlene Stephens, Anthony Turner, and Dennis Vaughan. In addition, I would like to thank Bruce Chandler, Penelope Gouk, David Bryden, and Rod and Madge Webster for their insightful comments and suggestions on the presentation and identification of the diptychs while they were on exhibit at the Houghton Rare Book Library, Harvard University, during the 1990 symposium of the International Scientific Instrument Commission. Finally, I wish to thank Nancy Finlay and Hugh Amory for assistance with obtaining prints from several rare books at the Houghton Library for inclusion in this catalogue.

Steven Lloyd
Thanksgiving Day, 1990
"SOLI DEO GLORIA"

1

The Use of Diptych Sundials

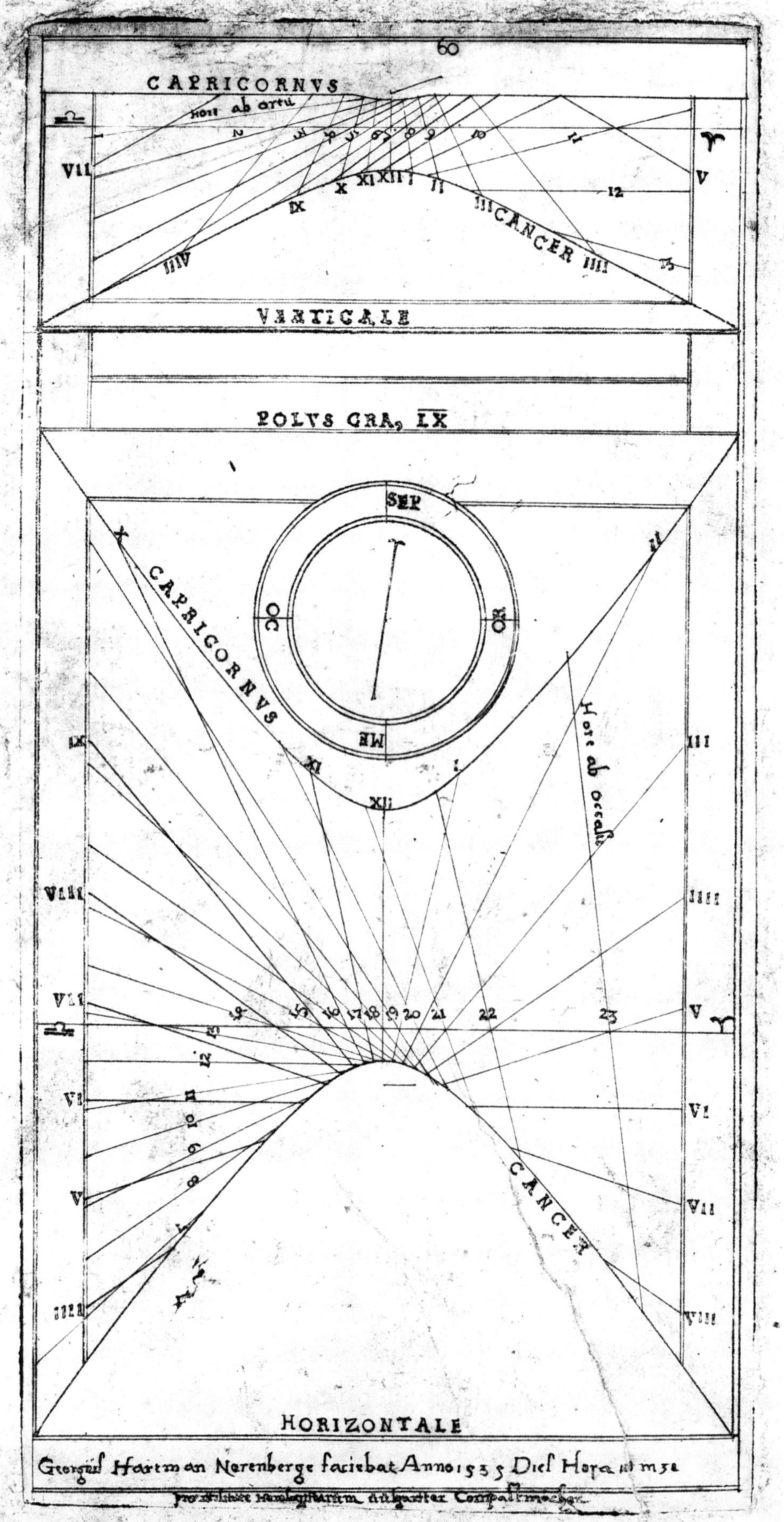

Georgius Hartman Norenberge faciebat Anno 1535 Diei Hora II m 30

Hour Systems

Unequal or Planetary Hours

The most common system of counting hours in the Middle Ages was to divide the periods of daylight and darkness into twelve hours each. Since the length of the day varies throughout the year, the daylight hours are longer or shorter than the nighttime hours, except at the two yearly equinoxes. Unequal hours were often called "temporal," "Jewish," or "planetary" hours (in Latin, *Planetarum Horae,* or in German, *Planeten Stunden*).

According to the principles of Renaissance astrology, the seven "planets" (Saturn, Jupiter, Mars, the sun, Venus, Mercury, and the moon) were thought to rule over the hours of the day in succession; in addition, the planet which rules the first hour of the day has a special astrological significance for the entire day.[1] For example, if we assign the first hour of the first day of the week to the sun, assign the second hour to Venus, and continue assigning consecutive hours to the planets in order of decreasing distance from the earth[2] according to the planetary system of Ptolemy (in the order listed above, repeating the cycle every seven hours), the first hour of the second day will be ruled by the moon. Hence in our present calendar we have Sunday followed by Monday (the moon's day). The names of the subsequent days of the week in English are derived from the transposition of the names of various Norse and Anglo-Saxon gods thought to correspond to the Roman pantheon, whose divinity in turn was associated with the seven planets.[3] Tuesday (from the Anglo-Saxon war god, Tiw) was thought to be ruled by Mars; Wednesday (from the Norse supreme god, Woden) by Mercury; Thursday (from the Norse thunder god, Thor) by Jupiter; Friday (from the Norse goddess of love and beauty, Freya) by Venus; and Saturday (from the Roman god of agriculture, Saturn) by Saturn. Occasionally one finds the assignment of planetary hours to the days of the week on the outer face of diptych sundials.[4] The determination of planetary hours was thought to be useful not only for astrological prognostication, but also for the practice of medicine. Renaissance astrology posited that heavenly occurrences — such as relative planetary positions or extraordinary events like the appearance of a comet — correspond to events in the sublunar world. Thus by knowing the planetary hours, medical practitioners could choose auspicious or "critical" times for administering medication or performing surgery.[5]

Unequal or temporal hours are indicated on diptych sundials by pin gnomon dials, usually located at the top of face Ib. A pin gnomon dial is simply a metal pin with hour lines inscribed around it. This type of

1. Bobinger (1966).

2. Gouk (1988), p. 19.

3. In the Romance languages, the names of the days of the week are similarly derived from the Greek and Roman pantheon.

4. For example, see Gouk (1988), plate XII (cat. no. 5, a dial by Hans Troschel the Younger, dated 1618).

5. See Garin (1983), pp. 93–94, and Siraisi (1990), pp. 135–136.

altitude dial is designed primarily to indicate the number of equal hours (see below) of daylight during each season. Such dials are typically labeled in Latin *Quantitas Diei* (see cat. no. 22) or in German *Tag Leng* (see cat. no. 5). In central Europe there are approximately eight hours of sunlight at the winter solstice and sixteen daylight hours of equivalent length at the summer solstice. When the dial is oriented properly, the length of the pin gnomon's shadow indicates the total number of hours of daylight. The daylight indicator can be used at any arbitrary time of day. Dials are often numbered 8 to 16 in one color (usually red) for the hours of daylight, and 16 to 8 in the reverse direction and in another color (usually black) for the corresponding length of the night. Zodiacal symbols are commonly used to indicate the months.

These pin gnomon dials sometimes include additional hour lines to indicate unequal-length temporal hours (see cat. nos. 13, 14, 28, 29, and 35). They are typically calibrated 0 or 1 to 12, corresponding to sunrise and sunset, respectively. Less often one encounters a separate pin gnomon dial calibrated solely for unequal planetary hours (see cat. no. 12).

6. Gouk (1988), p. 18. See Turner (1990), p. 21.

7. Vincent and Chandler (1969).

Equal Hours

During the fourteenth century, following the advent of the mechanical clock, equal hours came into use.[6] Although weight-driven mechanical clocks could not be easily adjusted to reckon unequal hours, which varied in length throughout the year, they were well suited to dividing the day and night into equal, reproducible units. These new mechanical devices were at first inaccurate and unreliable; the need to bring them into accord with local solar time as defined by the angular position of the sun in the sky led to the development of sundials calibrated to the same system of equal-length hours. For several centuries sundials provided a more reliable and accurate means of determining the time. In fact, the use of sundials actually increased significantly after the introduction of mechanical time-keeping. This was primarily due to the need to periodically adjust the mechanical clocks and bring them into correspondence with the local solar time, as well as to the increasing awareness of the utility of public and private timekeeping.

The diurnal period was divided either into twenty-four equal hours beginning at midnight or into two groups of twelve equal hours beginning at midnight and noon. The hours in a twenty-four-hour system were sometimes referred to as "large" or "great" hours (in German, *Grosse Uhr*), while the twelve-hour system counted "small" hours (*Kleine Uhr*, see cat. no. 3 for examples of both hour systems).

The system of small hours was apparently the dominant means of time reckoning in late sixteenth- and seventeenth-century Germany[7] and France, and it is still the most prevalent form of domestic (nonmilitary) timekeeping to this day. The primary horizontal dial (as well as the optional vertical dial) found on ivory diptychs is, with very few exceptions, *always* divided into small hours. This hour system is also referred to as "common" (*Horae Communes*), "German" (*Devtsch* or *Tevtsch*, see cat. no. 2), or "French" hours.

Three additional hour systems incorporating large hours persisted in use alongside the small or common hours throughout the sixteenth and seventeenth centuries. The inclusion of these other hour systems on the diptych dials may have been useful to merchants who traveled to foreign lands where time was reckoned differently. The most prevalent of these is the system of "Italian" hours (*Horae Italianae* or *H.I.*, see cat. no. 7), also known as "Bohemian" (*Behmisch* or *Pehmisch*) or "foreign" (*Welsch*, see cat. no. 2) hours. Italian hours are reckoned from sunset; dials indicating Italian hours are numbered only for the daylight hours, usually from about 10 to 24 (sunset). They are often labeled in Latin, *Horae ab Occasv [Solis]* (hours from sunset, see cat. no. 14), or in German, *die stund von Nidergang* (see cat. no. 11). Dials labeled with Italian hours and calibrated to the latitude of Venice are especially prominent among those made by the Ducher family workshop in Nuremberg (see in particular cat. no. 2); Venice was one of Nuremberg's primary mercantile trading partners throughout the sixteenth and early seventeenth centuries.

A second system of large hours is reckoned in "Babylonian" hours (*Horae Babyloniae* or *H.B.*, see cat. no. 7). Babylonian hours begin at sunrise and are calibrated from 1 to 16. They are often labeled in Latin, *Horae ab Ortv [Solis]* (hours from sunrise, see cat. no. 14), or in German, *die stund von Auffgang* (see cat. no. 11).

Both Italian and Babylonian hours are typically indicated by auxiliary pin gnomon dials located on faces Ib or IIa. While the two scales are sometimes placed side by side on separate pin gnomon dials (see cat. nos. 29 and 46), they are often found together on a single pin gnomon dial with two overlapping scales engraved in different colors (usually red and black, see cat. no. 19). Occasionally this combined scale is recessed into a shallow hemisphere with curved hour lines (see cat. no. 20); it is known as a concave pin gnomon or as a scaphe dial. In both the planar and scaphe dials, the shadow of the gnomon's tip indicates the local apparent time in the appropriate hour system. Hans and Thomas Ducher included a brass volvelle on a few of their dials for converting between common and either Italian or Babylonian hours (see cat. no. 1).[8]

A third system of large hours is actually a hybrid of the Italian and Babylonian systems: "Nuremberg" hours (*Horae Norimbergenses*) reckon the daylight hours according to the Babylonian system and night hours by the Italian system. This correspondence explains why dials calibrated for Babylonian hours are sometimes labeled as Nuremberg hours.

8. See also Gouk (1988), figure 87 (cat. no. 37, a dial by Thomas Ducher, before 1645).

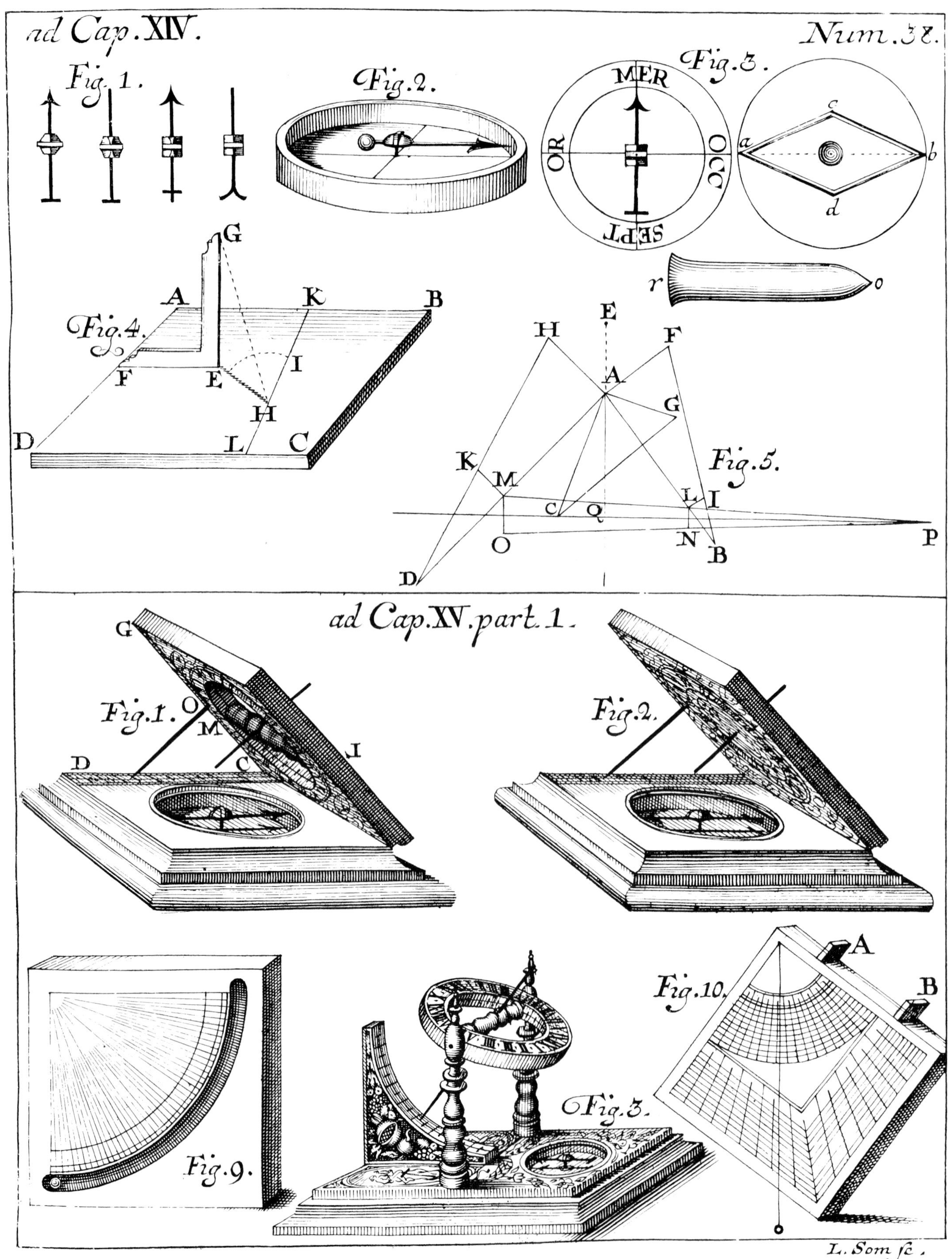
ad Cap. XIV.
Num. 38.
Fig. 1.
Fig. 2.
Fig. 3.
MER
OR
OCC
SEPT
a
b
c
d
r
o
Fig. 4.
G
A
K
B
F
E
I
D
H
L
C
H
E
F
A
G
K
M
L
I
c
Q
O
N
B
P
Fig. 5.
ad Cap. XV. part. 1.
G
Fig. 1.
O
M
D
C
I
Fig. 2.
Fig. 9.
Fig. 3.
Fig. 10.
A
B
L. Som fc.

Dial Furniture

String Gnomon Dials

With only a few notable exceptions (see cat. nos. 56, 81, and 82), the single distinctive element for any ivory diptych sundial is the inclusion of a string gnomon dial on face IIa. A string gnomon dial is simply a horizontal dial in which a string serves as the shadow-caster.[9] The string is inclined with respect to the horizontal dial plate at an angle equal to the latitude of the observer.

Many diptychs were designed for a single latitude (often either 48° or 49° for Nuremberg or 45° for Venice, see cat. nos. 2 and 46). Diptychs in which the angle of the string is fixed may also be paired with a vertical dial on face Ib; if the two tablets are held perpendicular to each other (often facilitated by small brass clasps which keep the two tablets fixed at a right angle), the angle that the string gnomon makes with the horizontal tablet will equal the latitude, and the angle made with the vertical tablet will equal the co-latitude (the complement of the latitude). While some multiple sundials are "self-orienting," meaning that the user is assured of having correctly oriented the dial by reading the *same* time on two separate dials, this is not true of diptychs; the horizontal and vertical dials will *always* indicate the same time, regardless of orientation.

Other diptychs are designed for use at a variety of latitudes. In order for these dials to indicate the correct time, some provision must be made for adjusting the angle which the string gnomon makes with the horizontal tablet. The most common method of providing for a variety of fixed latitudes was to drill a series of holes in the vertical tablet for different latitudes, through which the string gnomon could be attached. The string gnomon can be set by passing it through the appropriate hole and fixing it in place by means of a small wooden peg, or simply by tying a knot to hold the string taut. On Nuremberg diptychs, holes were commonly provided for 42, 45, 48, 51, and 54 degrees, corresponding to every third degree throughout Europe. These holes are often labeled *Elevatio Poli* (see cat. nos. 1 and 3) or *Polhoehe*, meaning the height or angular elevation of the pole star, Polaris, which corresponds to the latitude. Occasionally one finds additional unlabeled holes drilled in the vertical plate, indicating that the dial was modified at some later date for use at additional latitudes (see cat. no. 35).

Often a table of prominent cities and their latitudes was included either on face Ib or IIb on the Nuremberg diptychs, and on face(s) Ib and/or IIa on the French diptychs. Bloud-type diptychs manufactured in

9. Technically, the string is a "style" or shadow-casting edge, rather than a "gnomon," which is a rod or pole that marks the time by the length of its shadow. The term *string gnomon* has, however, gained general acceptance in the nomenclature of dialing and is adopted in this text.

Note the styles of compass needles (Fig. 1 at the top of the page) and compass bowl layout (Figs. 2 and 3). From Johannes Gaupp, *Gnomonica Mechanica Universalis, oder: die sehr deutlich und leicht vorgelegte allgemeine Mechanische Sonnen-Uhr-Kunst*, Augsburg, 1711. By permission of the Houghton Library, Harvard University.

Dieppe, France, usually include a printed or hand-written latitude list on the paper insert in the compass bowl. The lists typically included about two dozen locations, but some include up to a hundred cities.[10] Most of the cities listed on the Nuremberg dials are southern German cities, but well-known foreign capitals (such as London, Paris, Rome, and Constantinople) or major trading partners of the Nuremberg merchants (such as Venice, Milan, Danzig, Lübeck, and Antwerp) were also noted. Since the dials are calibrated to every third degree, the values of the latitudes are rounded up or down to match the "standard latitude zones" provided on the dial. While one might wish to know from what literary sources the dial-makers extracted their latitude tables (such as atlases and textbooks on geography or applied mathematics), this rounding-off to the nearest third whole degree renders comparison with contemporary printed lists of latitudes futile.

Another means of adjusting the angle of the string gnomon was a sliding metal scale on face Ib to which the top end of the string is attached. This method of adjustment was popular among the ivory diptychs produced in France. Latitude lists on these dials were not calibrated to the "latitude zones" as in the Nuremberg diptychs, since the angle of the string is continuously variable, usually from about 40 to 50 degrees (see cat. no. 53); they typically give the latitudes of cities to the nearest degree. Most of the cities listed on the French dials are located in France, although a handful of prominent foreign cities is occasionally included.

Horizontal dials designed for a variety of latitudes most often include separate, concentric sets of hour lines around the compass bowl (see cat. no. 18). Each set of hour lines corresponds to one of the latitudes or "latitude zones" to which the string gnomon can be set.

The use of the horizontal and vertical dials on a diptych is straightforward: Hold the dial in one hand and open the top tablet so that the two tablets are perpendicular, setting the metal clasps (if provided) to secure the vertical tablet. Hold the bottom tablet parallel to the ground and stand or sit with your back to the south, using the compass needle to find geographic north.[11] If the dial is designed for use at a variety of latitudes, fix the string gnomon in the appropriate hole in the vertical tablet. Make sure that the string remains taut when the dial is fully open. If the sun is shining, the shadow of the string gnomon on both the horizontal and vertical dials indicates the local apparent time. If several concentric scales are provided for the horizontal dial, read the time off the scale corresponding to the local "latitude zone."

A prominent exception to the sole use of common equal hours on the horizontal dial plate is a diptych probably made by Michael Lesel (see cat. no. 46). This instrument contains ten concentric scales on face IIb. The outermost scale is labeled 4–12–8 and indicates common hours for a single fixed latitude. The nine inner scales indicate Italian hours and are calibrated from 8–24 on the outermost (corresponding to sixteen daylight hours at the summer solstice) to 16–24 on the innermost scale (corresponding to eight daylight hours at the winter solstice). A table or almanac could be used to determine the appropriate scale to be used.

10. See, for example, Bryden (1988), cat. no. 69.

11. The magnetic declination between geographic and geomagnetic north is usually indicated in the compass bowl by an arrow: the dial will be geographically north-south when the compass needle lines up with this arrow. Occasionally one finds more than one line indicating magnetic declination engraved, scratched, or written on the bottom of the compass bowl or paper compass insert. As magnetic declination slowly shifted farther to the west throughout the seventeenth and eighteenth centuries, these new lines were added to indicate the current value of the magnetic declination. Additionally, paper compass bowl inserts might be rotated and glued into position to compensate for magnetic declination.

Magnetic Azimuth Dials

Many French ivory diptychs dating from the mid-to-late seventeenth century include an auxiliary magnetic azimuth dial in the compass bowl on face IIa. While there exist a few examples of magnetic azimuth dials prior to 1650, most dials of this variety are known as Bloud-type dials, after Charles Bloud of Dieppe, who introduced a very popular version of this sundial in the midcentury.

A facsimile of a pamphlet describing the use of Dieppe magnetic azimuth sundials (as well as their auxiliary equinoctial dials and lunar volvelles) is presented on pages 20–22 with an English translation. This rare pamphlet, housed in and reprinted with the permission of the Adler Planetarium in Chicago, provides a remarkably clear description of the use of such dials. While its author, "N.C.," is unknown, the date of the pamphlet corresponds to the year of perhaps the only dated dial by Charles Bloud, 1653 (see cat. no. 56).

The use of the Bloud-type magnetic azimuth dial is as follows. Turn the diptych over and set the moveable metal volvelle on the rear face (IIb) to the date. Rotating this volvelle moves the elliptical hour scale inside the compass bowl up and down[12] to compensate for the changing declination of the sun throughout the year. Now open the diptych, holding the lower tablet level, and orient yourself such that the shadow of the upper tablet completely covers the lower tablet. The position of the compass needle on the oval hour scale in the compass bowl will indicate the local apparent time.

One of the advantages of magnetic azimuth sundials over string gnomon dials touted in the Dieppe booklet is that they allow the user to determine the time even when the sun is not shining brightly enough to cast a distinct shadow. The user is directed to point the upper tablet toward the sun; even when the sun is behind a cloud or on a misty day, one can often estimate the position of the sun and thus determine the time.

If the elliptical hour scale is aligned perfectly north-south (parallel to the sides of the diptych), the magnetic azimuth dial will read correctly only when the local magnetic variation is near zero — that is, when the magnetic and geographic norths are aligned. During the mid-to-late seventeenth century magnetic and geographic north did indeed agree for much of western Europe. The migration of magnetic north in later decades rendered the subsequent use of these dials inaccurate.

Not all Bloud-type magnetic azimuth dials were constructed with the assumption that magnetic declination was zero or negligible. The large Bloud dial dated 1653 (see cat. no. 56) includes an elliptical hour scale inclined 5° east of north to compensate for magnetic declination.

Equinoctial Dials

Some diptych sundials include auxiliary equinoctial sundials on the outer (and sometimes also inner) faces of the upper tablet of ivory. The equi-

12. See page 116 for details.

Translation of *Usage de l'Orloge ou Cadran Azimuttal* by "N.C."

This pamphlet, *Usage de l'Orloge ou Cadran Azimuttal*, was found in the case of a Dieppe magnetic azimuth diptych sundial now in the Adler Planetarium. Previously, the dial and case were in the David P. Wheatland collection. The photographs of this rare pamphlet, courtesy of the Adler Planetarium, together with a translation are published here for the first time. Translation from the French by William R. Newman and A. J. Turner.

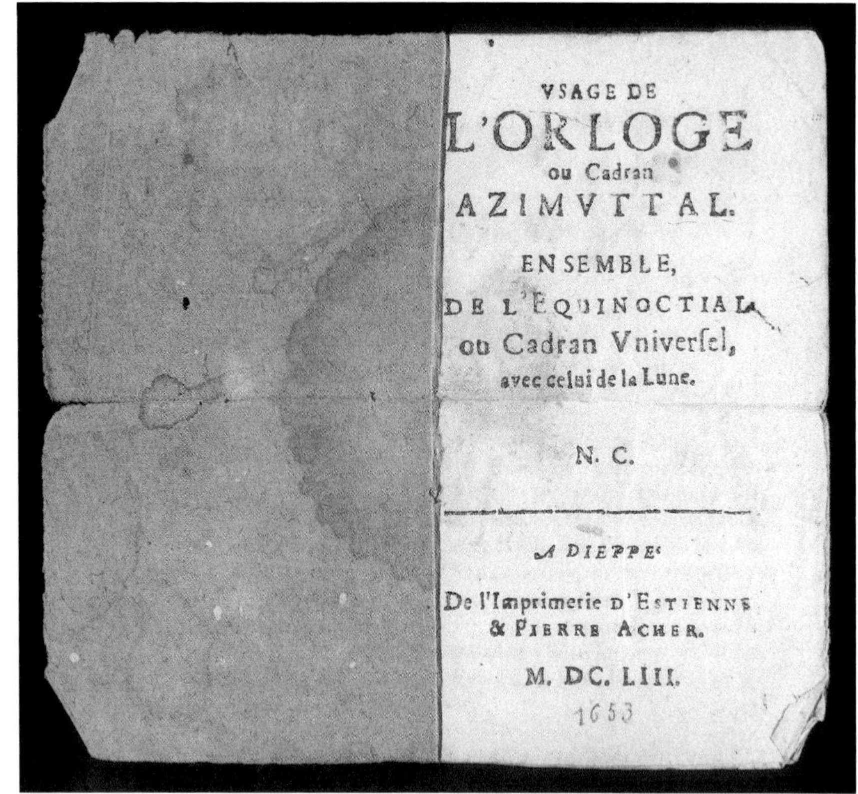

Use of
the Timepiece
or Azimuth Dial.
Combination
of the Equinoctial
or Universal Dial
with that of the Moon.

N.C.

At Dieppe

From the press of Estienne
& Pierre Acher.

1653

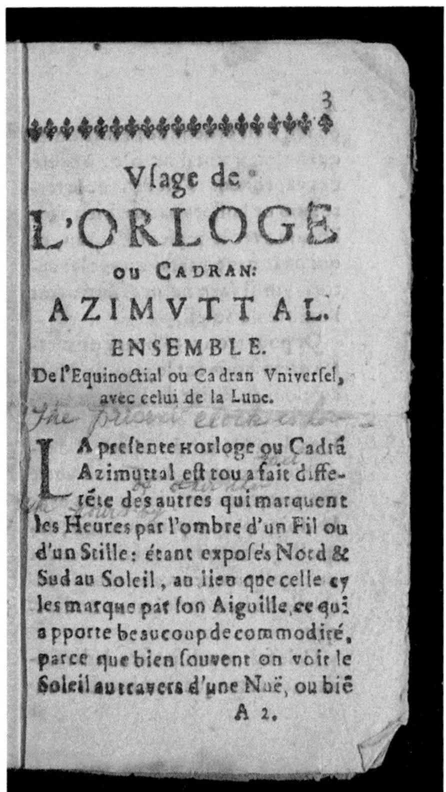

Vſage de
L'ORLOGE
ou Cadran:
AZIMVTTAL.
ENSEMBLE.
De l'Equinoctial ou Cadran Vniverſel,
avec celui de la Lune.

LA preſente Horloge ou Cadrã Azimuttal eſt tou a fait diffe-rête des autres qui marquent les Heures par l'ombre d'un Fil ou d'un Stille: étant expoſés Nord & Sud au Soleil, au lieu que celle cy les marque par ſon Aiguille, ce qui apporte beaucoup de commodité, parce que bien ſouvent on voit le Soleil au travers d'une Nuë, ou biê

A 2.

Use of
the Timepiece
or Azimuth Dial.
Combination
of the Equinoctial or Universal Dial,
with that of the Moon.

The present Timepiece or Azimuth Dial is quite different from others that tell the time by the shadow of a thread or a gnomon being placed North and South of the Sun: instead this dial indicates the hours by its compass needle, which adds much to its convenience because very often one sees the sun through a cloud, or again when it is near the horizon and makes no shadow because of the vapors that rise from the earth: one will nevertheless see the time clearly with the present dial; this cannot be done with other dials which need the sun's shadow.

Now to find the time one first aligns the day of the month with the tip of the finger of the hand shown on the bottom of the dial. The day of the month is inscribed on a piece of silver or other material which one can rotate with one's finger by a small knob. Turning this makes the hours advance or recede, according to the degree or place of the sun in the zodiac.

Once this is done, one must place the dial so that the shadow of the cover totally covers the interior of the dial, if the sun is casting a shadow. Whatever the case, it is always necessary for the exterior of the cover to face the sun, whether the Sun is casting a shadow or not, the needle, at rest, will indicate the time.

Use of the Equinoctial or Universal Dial

To find the time with this dial, one must know the latitude of the place where one wants to use it. Once this is established, the dial is inclined to the appropriate latitude against a support that is stored in a slot beside the needle. Then a gnomon, which is stored in a small hole covered by a small latch, is placed in the middle of the sundial. Then the dial is aligned North and South by its compass needle, and

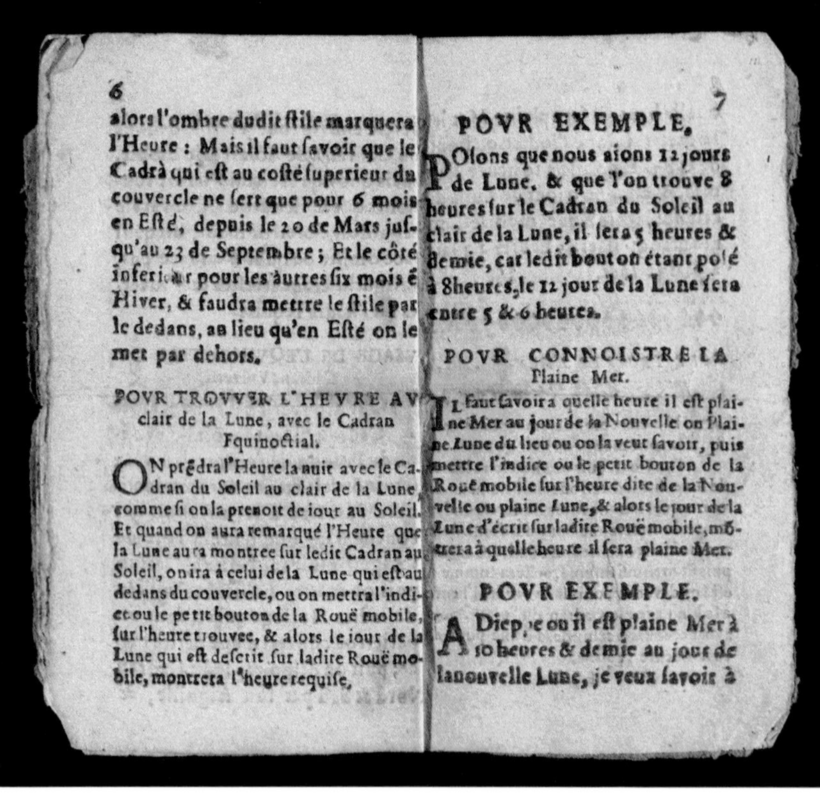

6

alors l'ombre dudit ſtile marquera
l'Heure : Mais il faut ſavoir que le
Cadrà qui eſt au coſté ſuperieur du
couvercle ne ſert que pour 6 mois
en Eſté, depuis le 20 de Mars juſ-
qu'au 23 de Septembre ; Et le côté
inferieur pour les autres ſix mois é
Hiver, & faudra mettre le ſtile par
le dedans, au lieu qu'en Eſté on le
met par dehors.

POVR TROVVER L'HEVRE AV
clair de la Lune, avec le Cadran
Fquinoctial.

ON prédra l'Heure la nuit avec le Cadran du Soleil au clair de la Lune,
comme ſi on la prenoit de iour au Soleil.
Et quand on aura remarqué l'Heure que
la Lune aura montree ſur ledit Cadran au
Soleil, on ira à celui de la Lune qui eſt au
dedans du couvercle, ou on mettra l'indice ou le petit bouton de la Rouë mobile,
ſur l'heure trouvee, & alors le iour de la
Lune qui eſt deſcrit ſur ladite Rouë mobile, montrera l'heure requiſe.

7

POVR EXEMPLE.

POſons que nous aions 12 iours
de Lune, & que l'on trouve 8
heures ſur le Cadran du Soleil au
clair de la Lune, il ſera 5 heures &
demie, car ledit bouton étant poſé
à 8 heures, le 12 iour de la Lune ſera
entre 5 & 6 heures.

POVR CONNOISTRE LA
Plaine Mer.

IL faut ſavoir a quelle heure il eſt plaine Mer au iour de la Nouvelle ou Plaine Lune du lieu ou on la veut ſavoir, puis
mettre l'indice ou le petit bouton de la
Rouë mobile ſur l'heure dite de la Nouvelle ou plaine Lune, & alors le iour de la
Lune d'écrit ſur ladite Rouë mobile, mõ-
trera à quelle heure il ſera plaine Mer.

POVR EXEMPLE.

A Dieppe ou il eſt plaine Mer à
10 heures & demie au iour de
la nouvelle Lune, je veux ſavoir à

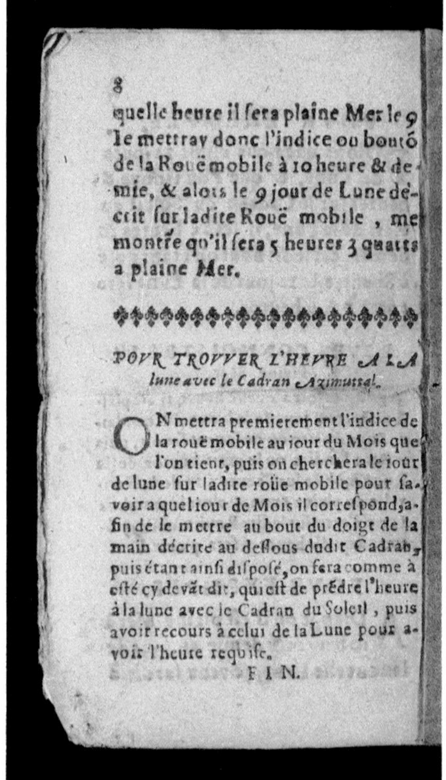

8

quelle heure il ſera plaine Mer le 9
le mettray donc l'indice ou boutõ
de la Rouë mobile à 10 heure & demie, & alors le 9 iour de Lune dé-
crit ſur ladite Rouë mobile, me
montre qu'il ſera 5 heures 3 quarts
a plaine Mer.

✤✤✤✤✤✤✤✤✤✤✤✤✤✤✤

POVR TROVVER L'HEVRE A LA
lune avec le Cadran Azimuttal.

ON mettra premierement l'indice de
la rouë mobile au iour du Mois que
l'on tient, puis on cherchera le iour
de lune ſur ladite roüe mobile pour ſavoir a quel iour de Mois il correſpond, a-
fin de le mettre au bout du doigt de la
main décrite au deſſous dudit Cadran,
puis étant ainſi diſpoſé, on fera comme à
eſté cy devãt dit, qui eſt de prédre l'heure
à la lune avec le Cadran du Soleil, puis
avoir recours à celui de la Lune pour a-
voir l'heure requiſe.

F I N.

---

the shadow of the gnomon will indicate the time. But it is necessary to know that the dial on the upper side of the cover works only for the 6 months of summer, from the 20 of March to the 23 of September; And the lower side is for the other 6 months in the winter. In winter, one must place the gnomon on the inside, while in summer one puts it on the outside.

To Find the Time by
Moonlight with the Equinoctial Dial

One finds the time at night with the sundial by the light of the Moon, just as if one were finding it by the Sun. And when one has noted the time indicated by moonlight on this dial, one refers to the moon dial [lunar volvelle], which is on the inside of the cover. There one places the index or the little knob of the moveable disc to the time found, and then the age of the moon that is inscribed on the same rotatable disc will show the required time.

For Example

Suppose that the moon is 12 days old and that, by moonlight, one finds 8 o'clock on the sundial, it will be 5:30, because the said knob being at 8:00, the 12th day of the moon will be between 5 and 6 o'clock.

To Find
High Tide

It is necessary to know the time of high tide on the day of the new or full moon at the place in question, then set the index or the little knob on the moveable disc to the time of the new or full moon. When this is done, the day of the moon marked on the moveable disc will indicate the time of high tide.

For Example

In Dieppe, where high tide occurs at 10:30 on the day of the new moon, I want to know

what time it will be high tide when the moon is 9 days old. Therefore I set the index of the moveable disc at 10:30, and then the 9th day of the moon inscribed on this disc will show me that high tide will occur at 5:45.

To Determine the Time
by the Moon with the Azimuth Dial

First set the index of the moveable disc to the present day of the month and then, on the same disc, find which day of the month corresponds with the current age of the moon. Align this with the tip of the finger of the hand inscribed on the underside of the dial. Once this is done, as explained earlier, find the time by moonlight with the sundial, and then use the lunar volvelle to determine the correct time.

THE END

noctial dial is perhaps the simplest form of equal-hour sundial. It can be used at any latitude, provided the upper tablet is inclined (with respect to the horizontal tablet) to an angle equal to the co-latitude of the observer (see figure 1) so that the dial face is parallel to the equator. This is accomplished by a metal index arm on the side of the upper or lower tablet; when set to the proper latitude position along a calibrated scale on the side of the upper or lower tablet, the upper tablet rests at the desired angle. A thin metal rod, which serves as the gnomon, is placed at the center of the circular dial plate on face Ia and sometimes also on face Ib. In an equinoctial sundial, the dial plate is parallel to the equator and the gnomon is parallel to the earth's axis. In such an arrangement the sun will shine on the upper dial face (Ia) when the sun's declination is above the equator (from March 21 until September 23) and on the lower dial face (Ib) for the remainder of the year. A single gnomon rod which extends both above and below the upper tablet may be employed. When not in use, the gnomon rod is stored in a small hole in the side of the lower tablet and is kept in place by a metal clasp. These thin rods are easily lost if the clasp comes loose; few dials have survived the intervening centuries with their gnomons intact. Harvard's collection includes only three intact gnomons, on cat. nos. 63, 70, and 76.

Both the upper and lower dial plates are divided into 15° increments, with 12 noon lying in the meridian and the 6 a.m. and 6 p.m. hour lines directed due west and east, respectively. The local apparent solar time is indicated by the shadow of the gnomon rod on the dial plate.

Harvard's collection contains only two Nuremberg diptychs that include equinoctial dials, made by Hans Troschel the Elder in 1611 and his son Hans Troschel the Younger in 1626 (see cat. nos. 12 and 13, respectively). In the latter instrument, the dial is labeled HOROLOGIUM AEQVINOCTIALE. Equinoctial dials are common on French ivory diptychs, and they are found on all Bloud-type magnetic azimuth dials in this collection. On those dials equipped with lunar/tidal volvelles on face Ib, the outer scale calibrated 1 to 12 twice can serve simultaneously as the inner equinoctial dial plate and the outer scale for the lunar/tidal volvelle. (See the booklet on the use of Dieppe diptychs on pages 20–22.)

Polar Dials

Polar dials, while extremely rare on German diptychs, are relatively common auxiliary dials on face Ia of French ivory diptychs. Like the equinoctial dial, the polar dial can be used at any latitude. However, the polar dial's orientation is effectively the converse of the equinoctial dial: the dial plate is parallel to the earth's axis, and the metal pin gnomon is parallel to the equator. To align the dial properly, rotate the dial 180° from standard usage so that the hinged edge is oriented toward the south. An index arm, often a thin pointer recessed in the bottom tablet, allows the user to set the angle between the ivory plates equal to the latitude of the observer. The local apparent time is indicated by the length of the gnomon's shadow on the dial plate. As with the equinoctial dials, the removable gnomon rod is stored in a clasped recess in the side of the bottom plate.

One must take care in adjusting the angle between the tablets on French diptych and Bloud-type magnetic azimuth dials. On these dials the equinoctial and polar dials are often combined, with a single gnomon serving both (see figure 1). Typically the polar dial is at the center, surrounded by a circular equinoctial dial calibrated (needlessly) for a complete twenty-four

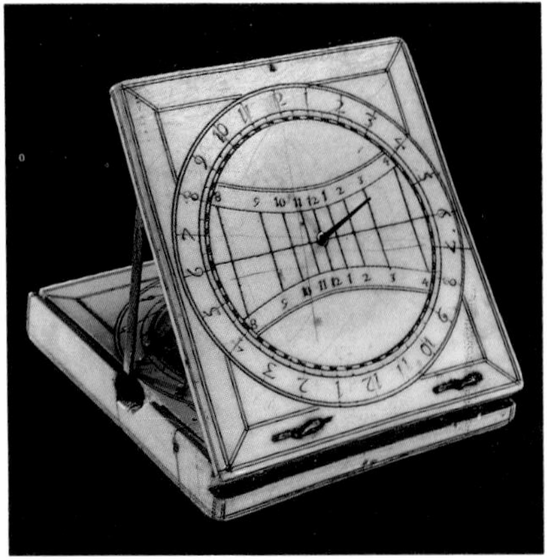

Figure 1   Dieppe ivory diptych sundial, showing its pin gnomon rod in position for use with the equinoctial and polar dials on face Ia. Catalogue no. 76.

hours. On these dials the scale against which the adjustment arm rests is calibrated to automatically yield the co-latitude. Therefore the scale can be used for the equinoctial dial "as is," but when using the polar dial one must set the adjustment arm to the co-latitude on this scale (such that the upper face will be inclined to an angle equal to the latitude); no doubt this mathematical double-negative was a source of confusion for the original users as well.

Volvelles

A very common feature of Nuremberg and French Dieppe diptych sundials is the inclusion of a lunar volvelle to convert the time at night as read off a sundial, from a shadow cast by moonlight, to the corresponding local solar time. A lunar volvelle is a simple calculating device for converting between lunar and solar time. Of the 82 ivory diptych sundials in Harvard's collection, over 80 per cent were constructed with some sort of lunar volvelle. Volvelles are a common feature of even the smallest "economy model" diptychs (such as cat. nos. 40–46).

The lunar volvelle found on most Nuremberg diptychs, "the standard German lunar volvelle," has three scales. The inner scale, on a rotating brass disc with a pointer, is labeled 1 to 12 twice. The two outer ones are fixed, engraved on the ivory. The first of these scales is labeled 1 to 12 twice and the outer 1 to 29 or 29½, with 1 corresponding to the new moon and 15 approximately corresponding to the full moon.

The use of the "standard German lunar volvelle" is as follows (see figure 2): Open the diptych, align it properly, and read the time off either the horizontal or vertical dials as indicated by the shadow of the string gnomon as cast by moonlight. Close the diptych and turn to the volvelle located on either the front or rear face of the instrument. Set the indicator or pointer of the rotating brass disc to the correct age or phase of the moon as indicated on the outermost scale. Once the volvelle is set, the "moon time" from the sundial is located on the scale on the rotating disc. The local apparent solar time is the corresponding time on the middle scale (on the ivory dial plate). It does not matter which half of either scale is used: a reading of 5 o'clock on either half of the inner scale will correspond to the same numerical value of the solar time on the middle scale.

Often the phase of the moon is written near the outer scale, either in words (usually in German) or pictures (circles indicating the relative size of the lunar crescent). Occasionally the brass volvelle has a circular hole, beneath which is a painted or incised crescent, so that when the volvelle is rotated the relative size of the lunar crescent appears through the hole. The volvelle is often decorated with a smiling crescent moon ("the man-in-the-moon") motif, indicating that this scale is intended for "moon time."

The time on the diptych dial corresponding to the lunar shadow, the "local moon time," will of course not be the same as the local solar time, except at the new moon. The lunar month is approximately 29½ solar days, which means that the moon takes about 29½ days to return to the same position in the sky with respect to the sun. The consequence of this is that the moon appears at the same position in the sky about 49 minutes later each evening. The volvelle is a simple analog device for calculating the accumulated correction factors between "local solar" and "local moon" times since the last new moon. Some volvelles include additional concentric scales which indicate the value of this correction, tabulated in hours[13] and minutes (see cat. no. 33). Such tables assume a correction of 48 minutes per day.

There are two conditions to the use of the lunar volvelle. First, the moon must be shining brightly at night. This limits use to approximately the second and third quarters of the lunar cycle. It is interesting to note that the glowing whiteness of the ivory in the moonlight greatly aids the observer in seeing the faint shadow cast by the moon. Second, one must know the age or phase of the moon. One can either count off the days on a calendar or simply estimate the phase by observation.

French ivory diptych dials often include a lunar/tidal volvelle on face Ib or, less commonly, on face Ia. Two varieties of lunar volvelles are found on the French dials. Both include an inner scale calibrated 1 to 29, 30, or 31 and an outer scale labeled 1–12 twice.

The first variety, which we call a "type 1 French lunar volvelle" (see figure 3), has two rotating discs and a fixed outer scale. The first (top) rotating disc has a single pointer and no scale. Usually there is a small hole through which the relative phase of the moon can be seen. The second

Figure 2　An example of a standard German lunar volvelle. Face Ia of catalogue no. 27.

13. Gouk (1988), figure 63 (a dial by Hans Troschel the Elder).

(bottom) rotating disc, of larger diameter, rests under the first. This disc has a pointer and on its outer edge the scale labeled 1 to 29, 30, or 31. The outer fixed scale, 1 to 12 twice, is engraved either on the ivory or on a third (larger) fixed metal disc. To find the time at night, set the pointer of the bottom rotating disc to the time, on the outer fixed scale, read off one of the diptych's sundials as cast by moonlight. Then, keeping this disc from moving, set the pointer of the top disc to the age of the moon on the scale of the bottom rotating disc. The pointer of the top disc will now indicate the corresponding local solar time on the outer fixed scale.

The second variety, which we call a "type 2 French lunar volvelle" (described in the booklet *Vsage de l'Orloge ou Cadran Azimvttal,* page 20 above, and see figure 4, below), has only one rotating disc, with a scale labeled 1 to 29, 30, or 31 and a knob or index located at the "1" position corresponding to the new moon (as in cat. no. 73). This rotating disc is surrounded by a fixed scale, labeled 1 to 12 twice, engraved on the ivory. To use the volvelle simply set the knob or index on the rotating disc against the time, on the fixed scale, read off one of the diptych's sundials as cast by moonlight. Now locate the age of the moon on the rotating scale; reading across to the outer fixed scale will yield the corresponding local solar time.

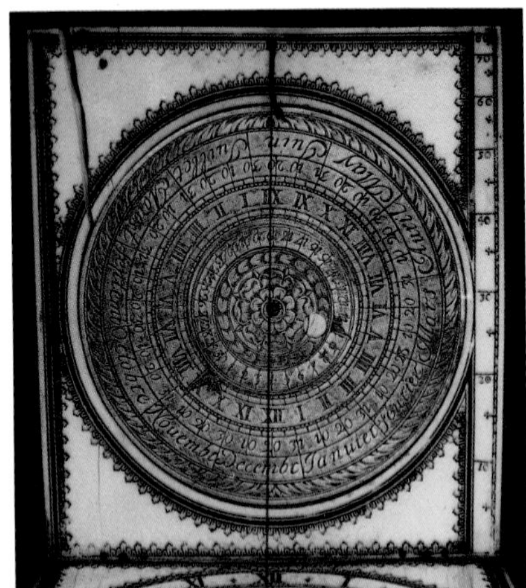

Figure 3   An example of a type 1 French lunar volvelle. Face Ib of catalogue no. 59.

Either of these French lunar/tidal volvelles can be used to determine the time of high tide for a given port. If one knows the time of the high (or low) tide for a given port on the full or new moon, one can set the indicator or the tiny knob at the "1" position on the metal volvelle to this time on the outer scale. Once set, simply read off the corresponding times for the high (or low) tides on the second, third, and subsequent days of the lunar cycle. As with the standard lunar volvelle, the high tides appear approximately 49 minutes later each day throughout the lunar cycle. Many French port cities are included on the latitude lists of French ivory diptychs, but the citizens of the largely landlocked German cities would have no use for a tidal volvelle.

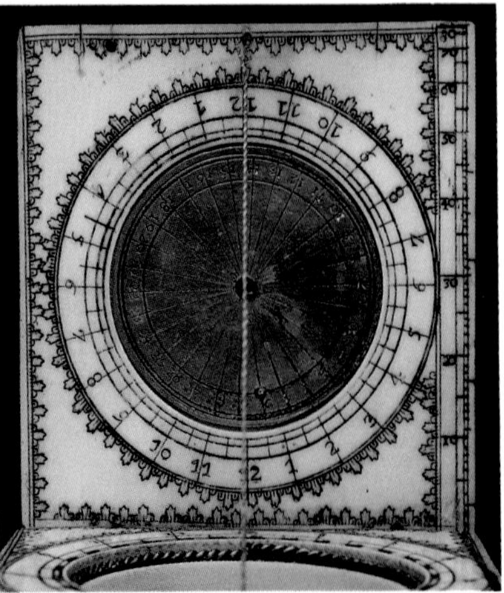

Figure 4   An example of a type 2 French lunar volvelle. Face Ib of catalogue no. 73.

Figure 5 Flemish ivory diptych sundial with a nocturnal. Catalogue no. 79.

Nocturnals

A few rare diptychs possess a device which enables the telling of time at night by the circumpolar stars. The single example of an ivory diptych with a nocturnal in Harvard's collection (see figure 5 and cat. no. 79) was probably made in Flanders and is dated 1599.[14]

The use of the nocturnal is straightforward. Set the midnight pointer of the brass disc to the date. Then hold the dial aloft and sight the pole star, Polaris, through the hole in the center of the dial. Now move the long index arm until you can sight along the edge of the Big Dipper to the stars Merak or Dubhe; or, if the nocturnal is calibrated for the Little Dipper, sight along the index arm to the star Kochab.[15] The time is read as the position of the long index arm along the disc's solar scale (labeled *HORA SOLIS*). A series of ratchet-like teeth along the edge of the brass disc (one tooth per hour) allows the user to *feel* the position of the index relative to the midnight pointer, and hence to find the time when there is little or no moonlight.

14. Compare with a very similar dial in the Bielefeld Collection dated 1598, shown in Syndram (1989), pp. 85–86.

15. Wynter and Turner (1975), p. 67.

16. See, for example, Bede's *De temporibus,* written in the year 703, in Jones (1943), p. 295. See also Dutka (1988).

17. See the article entitled "Calendar" in the *Encyclopaedia Brittanica,* 11th ed. (Cambridge, Eng., 1910).

Epact Tables

The problem of determining the proper date for the celebration of annual church holidays is almost as old as Christendom itself. The problem is essentially one of correlating the solar and lunar calendars to determine the dates for the moveable liturgical holidays (such as Easter and Pentecost, as determined by the lunar calendar) with reference to the fixed holidays (such as Christmas, as determined by the solar calendar). In 325 A.D. the Council of Nicaea, which also promulgated the doctrine of the Trinity, fixed the celebration of Easter on the first Sunday following the first full moon on or after the vernal equinox. Several numerical methods and mechanical devices were constructed as far back as the eighth century to facilitate the reckoning of church holidays.[16] One such mathematical construct was a system of golden numbers correlating the Julian calendar and the nineteen-year metonic cycle of lunar years.

After the imposition of the Gregorian calendar by Pope Gregory XIII in 1582, the system of golden numbers was replaced by a similar system of numbers known as epacts. An epact indicates the moon's age at the beginning of the calendar year, and thus aids in correlating the solar and lunar calendars. The sequence of epacts for the Gregorian calendar[17] over the entire nineteen-year period (beginning with the year 1600) is as follows: 15, 26, 7, 18, 29, 10, 21, 2, 13, 24, 5, 16, 27, 8, 19, 1, 12, 23, 4. This sequence simply repeats itself every nineteen years and is valid with respect to the Gregorian calendar from its inception in 1582 until the year 1699. However, much of Protestant Europe, for political and/or religious reasons, declined to adopt the calendrical reform immediately. The following sequence of epacts (again beginning in 1600) is valid for the Julian calendar in the sixteenth and seventeenth centuries: 25, 6, 17, 28, 9, 20, 1, 12, 23, 4, 15, 26, 7, 18, 29, 11, 22, 3, 14.

Epact tables commonly appear in a circular format surrounding the lunar volvelle. Typically two sets of epacts are provided, one for the Julian and one for the Gregorian calendars (see, for example, cat. no. 11).

While one can reliably date to within a year or two the manufacture of certain families' sundials (such as the Troschels') by determining the year which corresponds to the first entry of the epact table, other families of dial-makers (such as the Karner) almost invariably use the same cycle of epacts beginning with the same order on all dials. For example, a dial by Conrad Karner dated 1622 (see cat. no. 28) begins with the Julian epact sequence 23, 4, 15, etc. The proper sequence beginning in 1622 should be 28, 9, 20, ...; thus if one tried to date this dial by the starting point of the epact sequence, one would be off by five or fourteen years. Another diptych by Conrad Karner, this one dated 1630 (see cat. no. 29), also begins with the same sequence of epacts, either by design or by default. The dates found on diptychs by Thomas and Joseph Ducher similarly do not correspond to the initial values for the epacts; moreover, on their dials the terms *Gregorian* and *Julian* are mislabeled (reversed).

Perpetual Calendars

Many Bloud-type magnetic azimuth dials incorporate a perpetual calendar on the metal volvelle located on face IIb. The perpetual calendar is a square table with seven rows and seven columns. The first row has entries 5, 7, 4, 12, 6, 3, 11 and the second 2, 10, 0, 1 and 9 (not to be confused with "19"), 0, 0, 8. The non-zero numbers in these rows represent the months of the year, where 1 = March. The next five rows are labeled sequentially 1–31 and represent the days of the month. The perpetual calendar enables one, with knowledge of the dominical letter for the year, to locate the days of the week for any arbitrary month in any year. The perpetual calendar grid is usually surrounded by a circular calendrical scale, which is used to set the position of the hour scale for the magnetic azimuth sundial inside the diptych (see, for example, cat. no. 63).

Wind Roses and Wind Vanes

Many diptych sundials were designed not only as horological but also as meteorological instruments. The majority of the Nuremberg ivory diptychs in Harvard's collection include a wind rose to indicate the wind direction. For a given fixed location, wind direction is often a fairly

reliable first-order indicator of prevailing weather patterns. (See figure 6.)

The use of the wind rose is as follows. Close the diptych so that it lies flat, using the brass clasps if necessary. Remove the wind vane from its compartment in the side of the bottom plate and insert it into the hole at the center of the wind rose. (Note that the string gnomon cannot simultaneously be attached to this hole when the wind vane is in place.) Hold the dial horizontal and look down through the compass viewing-hole in the top plate. Orient the dial so that the compass needle aligns with the arrow beneath it. The dial should be sufficiently far away from your body so that you are not blocking the wind. The wind vane's flag spins freely and will orient itself with the prevailing winds. Now align the wind vane indicator (usually a brass arm in the shape of a hand or an arrow) in the same direction and read off the wind's direction from the scale beneath.

Wind roses are typically divided into 16 or 32 directions. When so labeled, the numbers usually start at east and run in a clockwise fashion. While the directions on the Nuremberg dials are most commonly written in German (*Nort, Ost, Sud,* and *West*), one occasionally finds the Latin equivalents (*Septentrio, Occasus, Meridies,* and *Ortus*) inscribed on the wind rose. Several Miller dials include an interesting notation for some of the intercardinal points (see, for example, cat. no. 21); the sequence of points from north to south reads *Nort, Nort-Ost, Nort-Ost, Ost-Ost, Ost, Ost-Ost, Ost-Sud, Ost-Sud,* and *Sud.* In this naming scheme, several directions are ambiguous; for example, east-south-east and east-north-east are both represented by *Ost-Ost.* Several of the dials in Harvard's collection, especially those of Leonhart Miller, indicate either the names of the eight Mediterranean winds in Italian (*Ostro, Lebechio, Ponente, Mastro, Tramontan, Greco, Levante, Sirocho;* see cat. no. 20) or the corresponding anticipated weather in German (*Warm Feicht, Rengisch, Kalt Feicht, Schneig, Schon Drvcken, Heiter Kalt, Warm Heiter, Schon Mitelmesig;* see cat. no. 16).

Although most dials originally had a wind vane, very few original wind vanes exist today (see cat. no. 34). Similarly, many dials in this collection have lost their brass wind rose indicators, lunar volvelles, pin gnomons, or other accessory parts over the intervening centuries.

2

Nuremberg Diptych Sundials

Historical Introduction

by Penelope Gouk

Exactly when a magnetic compass was first added to a sundial has not yet been firmly established, although it is likely to have been before the fifteenth century.[1] The combination of sundial and compass in portable form was already known as a "compass" in the 1480s in Nuremberg, and by the early decades of the sixteenth century the "compass-maker," working in ivory, wood, and brass, had become a recognized specialist craftsman in the city. In idiomatic German of the sixteenth and seventeenth centuries, "compass" was synonymous with "sundial," as can be seen from works by Peter Apian and Sebastian Münster.[2]

Nuremberg was the first city in Europe to have an organized trade of compass-makers. The earliest use of the term *compass-maker* as a profession in the city records dates from the mid-1480s, and the rules of the compass-makers were included in the "book of Handicrafts" of 1535, the first codification of all such rules to be produced in the city.[3] Over the next century members of the craft gradually enforced strict regulations designed to protect themselves from outside competition. To a large extent they succeeded, but at the expense of any further innovation or development of the instruments they produced. The past splendor of Nuremberg handicrafts, including the work of the compass-makers, is appropriately commemorated in a manuscript of 1719:

> The craft of the compass-makers is of great reputation, because they made several types of compasses from wood, ivory, and brass, in various forms, such as in the form of little books, lutes, violins, rings, and many other types. These have been occasionally exported, even to distant-lying provinces and kingdoms. They also made quadrants in various ways for all types of hours, such as the German, the Bohemian, as well as ship and night compasses. But now there are fewer masters than before, and there is also less work, so that they can do very little about the decline of their craft. It is a closed craft, in that they are not allowed to travel outside in order to work in the same, otherwise they would be acting dishonestly and breaking their oath. The masterpieces for a sworn-in master consist of three dozen boxwood compasses of various sizes, also of three punches, whereby the path to the hour number is drawn onto the compasses, and then they have to design a quadrant freehand.[4]

It is evident that the craft at the beginning of the eighteenth century had noticeably declined from its former state.

Although a greater number of ivory diptych sundials survives than any other kind of instrument made by the Nuremberg compass-makers, these instruments were only one part of the range that was produced in the city.

1. An example of an astronomical compendium with a compass needle was made in Damascus in 1365/6 by ʿAlī ibn Ibrāhīm ibn al-Shāṭir; Janin and King (1977).

2. Wagner (1901).

3. Gouk (1988), pp. 77–81; see also Jegel (1965).

4. Original in Nürnberg Stadtarchiv, *Rugamt*, Rep. B. 12, vol. 236; transcribed in Gouk (1988), p. 81.

Sundial-maker at work. From a work signed B. L[eeman?], *Instrumentum Instrumentorum: Horologiorum Sciotericorum*, Zurich, 1604. By permission of the Houghton Library, Harvard University.

According to the craft rules, compass dials were to be made in ivory, good-quality boxwood, or pear wood. Poor-quality wood might be used only for "turned little boxes." Small wooden instruments were probably the mainstay of the compass-makers' trade, although because of their simplicity and perishability few examples have come down to us. Two simple, horizontal compass dials and part of a third now in the collection of the Museum of London have been found in archeological digs at Nonsuch Palace near Epsom and Worship Street, London EC2, and ten more thought to be of Nuremberg provenance have been recovered from the wreck of the *Mary Rose.*[5] The Germanisches Nationalmuseum in Nuremberg has four sixteenth-century wooden diptychs: two attributed to Erhard Etzlaub dated 1511 and 1513, one by Ulrich Marsch (before 1543), and one by Georg Reinmann dated 1555.[6] The Harvard collection includes thirteen examples of European wooden diptychs that were made at any time between the late seventeenth and early nineteenth centuries.

There are also eighteen examples of wood-and-paper diptychs, four of them by David Beringer (fl. 1777–1821),[7] in the Harvard Collection. Wood-and-paper compasses were already being made in sixteenth-century Nuremberg, although no examples that I am aware of survive. Several designs for instruments, including diptychs, published by Georg Hartmann between 1527 and 1563 still survive.[8] It is perhaps no coincidence that by 1565 complaints were already being made by compass-maker masters that cheap, shoddy instruments were being produced by some individuals from painted paper stuck onto wood, and a penalty for producing instruments by such means was added to the statutes in 1574.[9]

Compass-making was firmly placed among the wood and turning trades in Nuremberg, despite the fact that the instruments required at least some brass components. This was in contrast to Augsburg, where compasses and other scientific instruments were produced in brass and other metals by members of metal-working guilds, such as locksmiths, clock-makers, and gunsmiths.[10] However, brass diptychs signed by Nuremberg makers include examples by Hans Gruber (fl. 1552), Hans Friessfeldt (1555), Paul Reinmann (1601), and Hans Troschel (1609). Certain masters, notably Paul Reinmann and Hans and Thomas Ducher, used gilt brass extensively for the decoration of their ivory work. One example by Hans Ducher in the Museum of the History of Science, Oxford, has an upper leaf of gilt brass and a lower leaf of ivory with a wood core.[11]

The other relatively under-represented category among surviving Nuremberg instruments that deserves mention is the "form" or novelty compass, instruments in the form of little books, violins, rings, and many other types. Two examples, one in the form of a tiny ivory mandolin, the other a human skull, are in the Bayerisches Nationalmuseum, Munich. It is only by a tremendous effort of imagination that we can begin to grasp the size and diversity of the Nuremberg compass trade as a whole, which was itself only one (albeit major) part of European production of portable compass sundials in the sixteenth and seventeenth centuries.

Hundreds of Nuremberg ivory diptychs are found in museums and private collections throughout the world. On many of these instruments

5. Gouk (1988), pp. 28–29.

6. Zinner (1956; reprint, 1979), pp. 310, 438, 484. A 1513 Etzlaub diptych was obtained from the Wheatland collection by the Adler Planetarium.

7. Chandler and Vincent (1969).

8. Some of the originals from Munich, Bayerische Staatsbibliothek MS Rar. 434, are reproduced in Gouk (1988), pp. 68–69, 76, 105–106. See also page 12 above.

9. Gouk (1988), p. 79.

10. Bobinger (1966), p. 78.

11. Gouk (1988), pp. 65, 119; Zinner (1956; reprint, 1979), pp. 316, 327, 487, 552–553.

there is one or more of the following pieces of information: the maker's full name or initials; an identifying mark; the date of manufacture; the mark "N" or the name of the city. From the examples that I know (which one can only assume are representative of the trade as a whole), the following observations can be made. With relatively few exceptions all these surviving instruments were produced by members of six families between about 1550 and 1730: Troschel, Ducher, Karner, Lesel, Miller, and Reinmann. There are a handful of instruments signed by other known compass-makers: single examples by Lienhart Gresel (1531), Hans Felt (1566), and Georg Riege (1699), and several by Johann Gebhart (fl. 1538–1548). There are also some instruments signed (but probably not made) by the mathematicians Georg Hartmann (1489–1564) and Christian Heiden (1526–1576).[12] Four unsigned ivory diptychs of similar design (one complete and three incomplete) have been provisionally dated to the last decades of the fifteenth century; if this is correct, and indeed if they are from Nuremberg (which is not necessarily the case), they would be the earliest known examples.[13]

Most of the largest diptychs, elaborately decorated and of high quality, date from between about 1580 and 1620, while most of the smallest instruments, made from a "sandwich" of ivory (sometimes bone) with a wooden core, date from the later seventeenth and early eighteenth centuries. The decorations on these dials are usually engraved whereas the numbers and letters are usually punched. More detailed observations about the different styles of instruments produced by each of the family workshops are given below. The Harvard collection includes a representative sample of instruments produced between about 1575 and 1730, with examples from all family workshops except the Reinmanns. Although it does not have any exceptionally large German pieces, the collection is particularly strong in the "sandwich" dials which represent the end of the tradition for ivory.

Why did Nuremberg become a center for the manufacture of sundials, and why were ivory and wood the materials preferred for their production?[14] The production of such instruments required the collaboration of mathematicians and skilled craftsmen, suitable raw materials, and a means for the distribution of the finished goods. The city of Nuremberg, already an important center for the manufacture and distribution of luxury items, clearly met such requirements. Around 1500 about 150 different crafts were practiced in the city. The production of metal wares was particularly important, together with the related skills of die- and punch-cutting and engraving. The material resources for the production and marketing of compass sundials on a commercial scale were therefore readily available.

Manufacturing interests in the city were complemented by those of banking and commerce. Like the more famous Fuggers of Augsburg, a number of Nuremberg merchant families had a network of credit and trading facilities that extended throughout Europe and the New World. These merchant families used a system of trading associations, which involved relatives working in each foreign city. For example, around 1500 there were 232 Nuremberg merchants in Venice. The *Handel Buch* of the merchant Lorenz Meder published in 1558 shows how his business em-

12. Information on these are found in Gouk (1988), pp. 49, 59, 61, 64, 82, 93, 95, 110, 125, 133; and Zinner (1956; reprint, 1979), pp. 314, 319, 326, 364, 365, 367, 369–371, 491.

13. The British Museum has a complete example. Two upper leaves are also known, one in the Metropolitan Museum of Art, New York, the other in the museum at Niort, France; a lower leaf is in the Museum of the History of Science, Oxford; Gouk (1988), pp. 28–29.

14. These questions are addressed more fully in Gouk (1988), chaps. 3 and 4. For general background see Strauss (1966).

braced Lisbon, Lyon, Antwerp, Cracow, and Warsaw and was linked to all parts of the globe. In this trading account Meder records that in Lisbon one hundredweight of elephant tusks cost 1,170 pfennigs, while in Antwerp the same amount would have cost 1,320 pfennigs.[15] The ivory came from the west and east coasts of Africa via the new Atlantic trade routes recently opened up by the Portuguese. The traditional overland trade routes across Africa and their control by the Arabs were completely disrupted by this development. Ivory flooded into Europe in much larger quantities than had ever been previously available once the African traders dealt directly with the Europeans on the western and eastern coasts; their goods were imported to the Atlantic ports of Lisbon, Antwerp, and later Amsterdam. From Lisbon the ivory would have been transported by land and water across Spain and France, or else from Antwerp up the Rhine and the Main. The Nuremberg craftsmen would probably have bought their supplies from merchants and traders in the city itself.

As one of the free imperial cities in the empire, Nuremberg was a regular venue for imperial Diets or parliaments. The city's support for the emperor and later political ambiguity were critical to its growth and later decline. The imperial crown jewels had been transferred to Nuremberg in 1424, and the imperial seal was regularly deposited there while the emperor was absent from Germany. The protection of the emperor meant that Nuremberg escaped becoming dominated by a local prince, the fate of many other towns, and was able to extend its possessions and interests. By 1500 the city held extensive territories outside its walls; its control over lands and inhabitants extended over a radius of twenty-five miles. Support for the emperor became a liability during the Reformation, however. The city long refused to take direct action in the struggles between the Protestant alliance of princes and cities on the one hand and the emperor and the Catholic league on the other. It was the only Protestant city not to join the Schmalkaldic league of 1535. Nuremberg's attempt to maintain a neutral stance in the struggles of the Reformation led to a progressive decline of its political position, which eventually undermined the city's trade and manufacture.[16]

The unusual social structure of the city must also be noted, since it differed significantly from that found elsewhere. In the year 1348 a revolt by the craftsmen of the city against the patricians had been decisively overthrown. All guilds were abolished in the following year. From this point on only members of the patrician families exercised any significant power in the city, making them the greatest urban aristocracy in Germany. Councilors from these families controlled courts and legal procedures, set taxes, and made rules concerning public order and all aspects of civic life. The guilds had no political power to speak of in the city. It seems as though Lutheranism, adopted by the city in 1525, did not affect this trend of "benevolent despotism." At first it worked in favor of the city's trade and industry, but gradually it led to rigidity and fear of innovation, as is shown in the case of the compass-makers.

In 1500 Nuremberg numbered about twenty thousand people, with another twenty thousand living in close proximity, making it one of the largest population centers in Europe.[17] The upper class consisted of forty-

15. Kellenbenz (1974); for further details of the changing pattern of trade, see Parry (1963) and Bovill (1969).

16. Baron (1937); Seebass (1972).

17. Endres (1970).

two ancient families who provided thirty-four out of forty-two members of the city council. The remaining eight members were selected from representatives of the most important crafts, notably the building trades. Below the patricians, there were around four hundred "Honorable Families"; these occupied city offices and lower council posts, but they were effectively excluded from real political power. Together with the patricians they made up 6 to 8 per cent of the population. There was a substantial intermediate class of master craftsmen and lesser tradesmen in the city. In the mid-sixteenth century there were over five thousand masters of crafts. Their work was subject to extremely strict regulation and control by the council. City employees, laborers, and journeymen, most of whom were not citizens, constituted a further group. Together with the regularly unemployed, they made up a substantial underclass.

Nuremberg artisans were fortunate to have raw materials from the whole world at their disposal, as well as an international market for their wares. Because of this large market, the craftsmen were in a position to become specialists, highly trained individuals whose range of products was both limited and protected by strict council legislation. This legislation was also intended to protect technological secrets from outside competitors. Reflecting this concern with secrecy, there were basically two types of craft in Nuremberg, *Geschworne Handwerke* (sworn crafts) and *Freie Künste* (free crafts).[18] The closed or sworn crafts included those on which the prosperity of the citizens depended, notably those which used metals such as brass and gold. In contrast, the free crafts could be practiced by individuals without such control, and included artisans such as painters and woodcarvers. In practice some crafts, including the compass-makers, fell between the two types, which suggests that the distinction was less rigid than the statutes imply.

The development of Nuremberg's commerce and trade enabled cultural and intellectual life to flourish. The city was an ideal, but by no means unique, location for fruitful collaboration between scholars and craftsmen. As early as 1444 several astronomical instruments had been made by Nuremberg brass workers under the direction of the astronomer Nicholas Heybech (fl. 1425–1446).[19] The presence of craftsmen skilled in metalwork and engraving in Nuremberg acted as a powerful lure to the astronomer Johann Müller of Königsberg, better known as Regiomontanus (1436–1476).[20] In the year 1471 he set up a workshop and printing press in Nuremberg, with the intention of producing scientific instruments and books, a short-lived scheme which ended with his death in 1476. Regiomontanus is known to have been involved in the production of scientific instruments, though nothing has survived from his Nuremberg period; but it is a powerful coincidence that the production of compass sundials emerged as a new trade in the city on the heels of his arrival.

The earliest named compass-makers, Conrad Cristel, Ludwig Gribner, Jorg Bynter, and Wolfgang Pirger, come into the records in 1484–1485.[21] At just this time the multi-talented Erhard Etzlaub was granted the rights of a citizen in Nuremberg. Etzlaub's long career (he died in 1532) exemplifies the links between the production of scientific texts, maps, globes, and mathematical instruments in sixteenth-century Nuremberg.[22] He was

18. Gouk (1988), pp. 45–46; Pilz (1954), pp. 8–9; Stockbauer (1879).

19. Zinner (1956; reprint, 1979), pp. 286–287; Hartmann (1919).

20. Zinner (1990).

21. Gouk (1988), p. 64; Nuremberg Staatsarchiv, *Amts -u. Standbücher* Rep. 52b, vol. 305, fols. 186, 187, 190.

22. Gouk (1988), pp. 49, 50, 64–65, 86–88, 110; Schnelbögl (1966).

Location of Houses of Compass-makers in Nuremberg

1 **Weissgerbergasse** Hans Miller III lived here in 1623.

2 **Nägeleinsgasse** Jorg Miller lived here in 1564.

3 **Kreuzgasse** Hans Troschel lived here in 1582.

4 **Zwischen den Fleischbänken** Paul Reinmann died here in 1608 or 1609.

5 **Unter den Huttern** Hieronymus Reinmann died here in 1577; Paul Reinmann lived here in 1604 but later moved.

6 **Rossmarkt** Albrecht Karner died here in 1687; Hans Karner died here in 1716.

7 **Köthgasse** Erasmus Karner died here in 1651.

8 **Schmiedgasse** Jacob Karner lived here before 1648; Jacob Friedrich Karner lived here near Spittler Thor in 1686; Melchior Karner died here near St. Jacob's in 1707.

9 **Hohes Pflaster** Georg Reinmann II lived here in 1571; Georg Reinmann III lived here in 1575; Albrecht Karner bought a house in 1646; Georg Karner lived near here before 1684.

10 **Schottengasse** Caspar Karner bought a house here in 1579.

11 **Engelhardsgasse** Christoph Ducher died here in 1632.

12 **Pfaffengässlein** Nicolaus Miller had a house here in 1647.

13 **Maiengasse** Hans Ducher III lived here in 1570; Hans Karner I sold a property here in 1610; Hans Paulus Karner lived here in 1626; Hans Reinmann II died here in 1627; Hans Reinmann III died here in 1628.

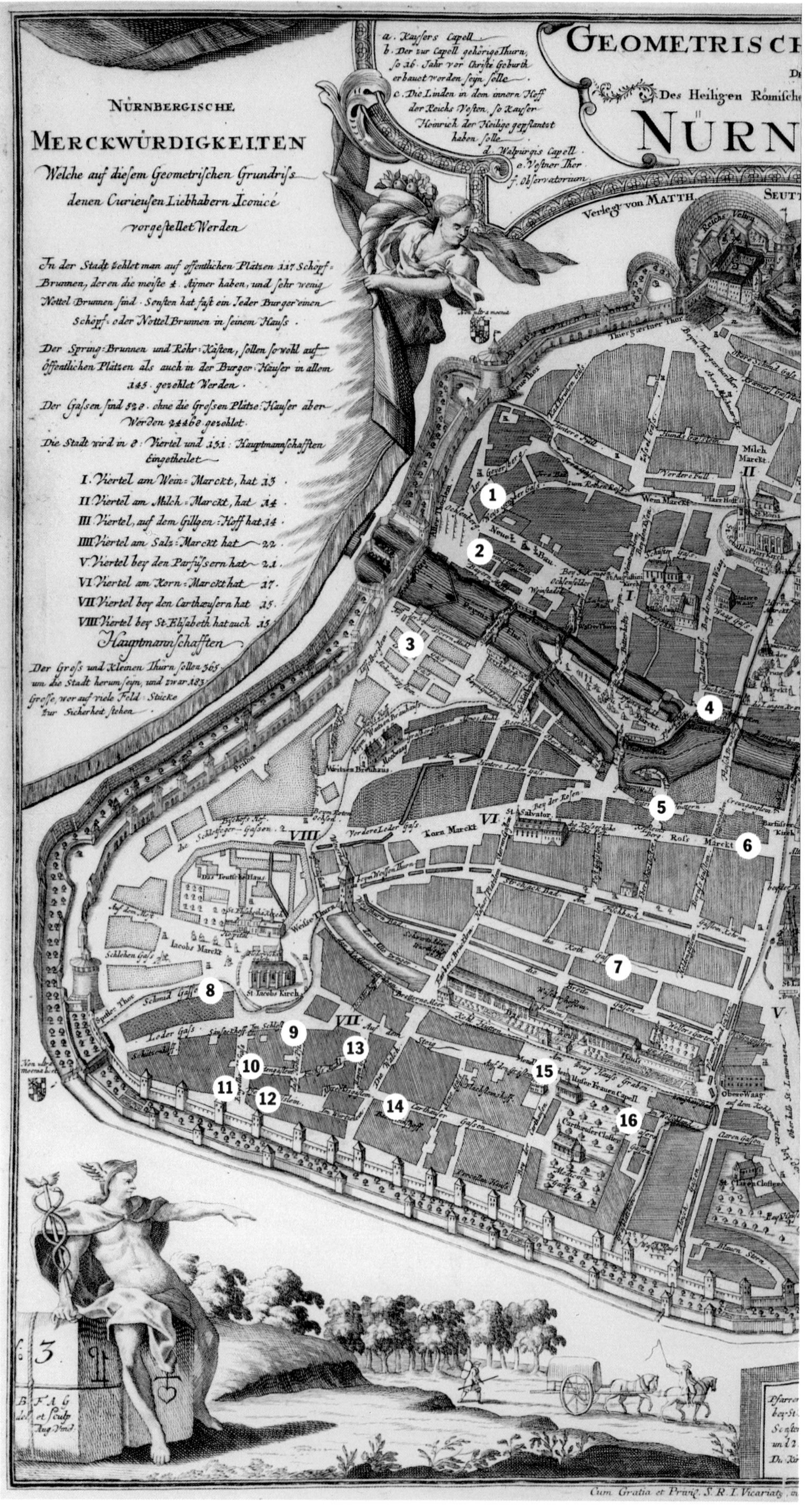

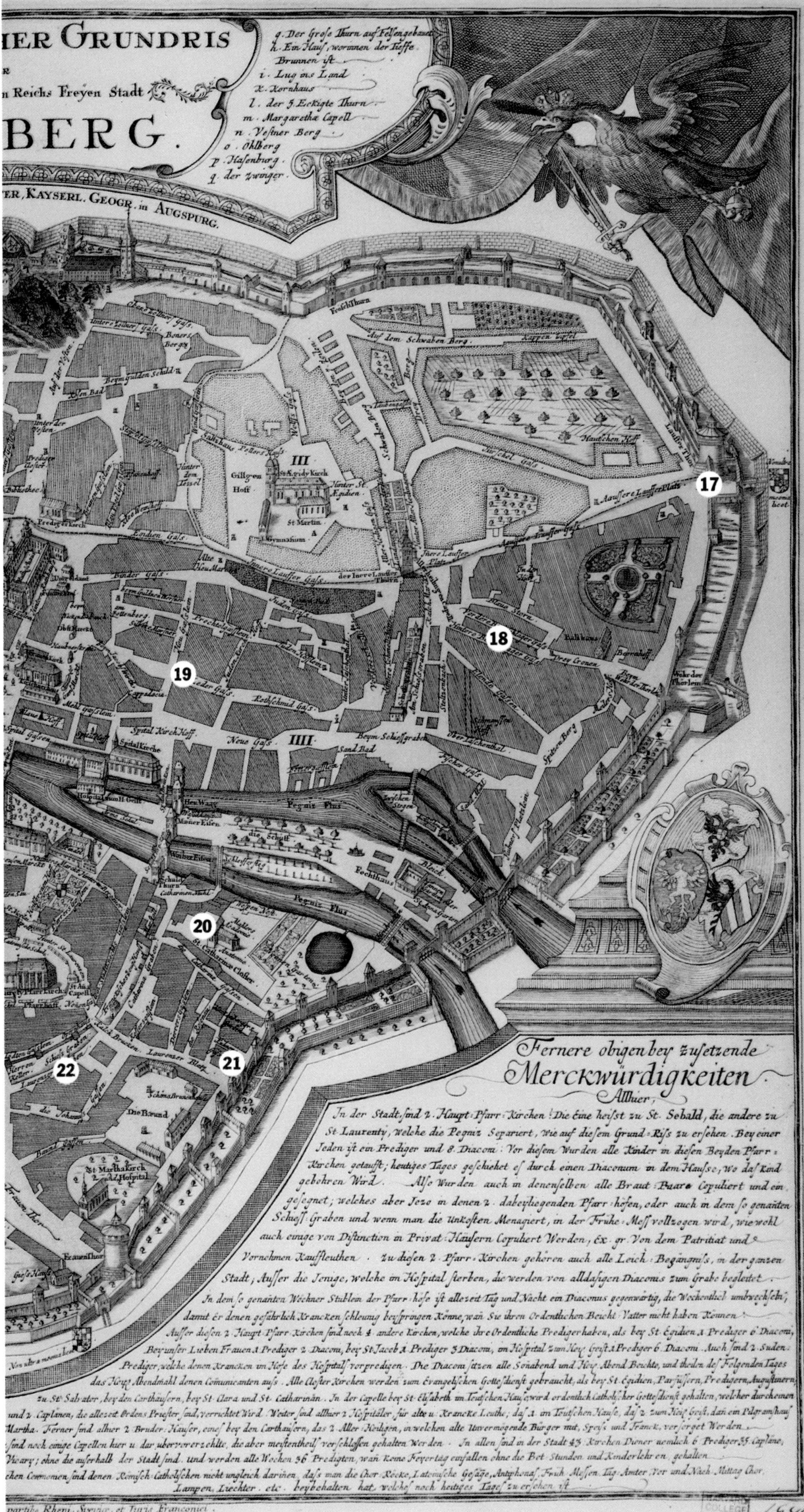

14 **Kartäusergasse** Hans Karner I
lived here in 1615; Conrad
Karner II lived here 1627-32;
Hans Karner II was a landlord
here 1645-74.

15 **Mendelzwölfbrüderhaus** Hans
Ducher III lived his last years
here (1632).

16 **Grasergasse** Hans Ducher II
lived here 1567-1615; Hans
Troschel bought a house here
in 1618.

17 **Am Laufer Schlagturm** Erhard
Etzlaub lived here in the 1500's,
by the old smelting house.

18 **Hintere Beckschlagergasse**
Leonhart Miller lived here
1637-41.

19 **Ledergasse (Alte Ledergasse?)**
Hans Troschel the Younger
bought a house here in 1620.

20 **Am Katharinen Kloster**
Properties on this street were
sold by Hans Ducher II (1611)
and Hans Karner I (1612), and
a house was sold by Hans
Karner II (1634).

21 **Kühnertsgasse** Hans Karner I
had a property here in 1610;
Andreas Karner died here in
1682.

22 **Lorenzergraben** Thomas Ducher
died here in 1645.

Not located on map:
Auf dem Platz Hans Ducher I
died here in 1550.

Auf der Walsch Hans Miller I
died here in 1555.

Paniersberg Michael Scheltner
had a house here in 1534.

Pfeifergasse Hans Reinmann I
died here in 1571.

at various times a surveyor, map-maker, medical practitioner, and almanac-maker. In 1507 one brother of the celebrated Martin Behaim wrote to another brother concerning the delivery to Lisbon of compasses being made by Etzlaub. According to Johannes Cochlaeus in 1512, Etzlaub's instruments were sought after "even in Rome."

In about 1500 Etzlaub published the "Rom Weg," the oldest surviving published map of Germany and central Europe, which was designed for use with a compass dial. Instructions how to use the map (oriented with south at the top) are included along with a representation of a sundial with a south-pointing needle. The wooden dials of 1511 and 1513 attributed to Etzlaub have string-gnomon dials adjustable for a series of latitudes, as well as a map of Europe and the Mediterranean oriented from south to north together with a latitude scale on the outside of the upper leaf. Although the adjustable gnomon is a feature of many ivory diptychs produced later in the century, the method of representing latitudes by means of a map is not; such information, if presented at all, is in tabular form, which would have been easier to reproduce. Where the compass-makers obtained their lists cannot be determined, but during the course of the sixteenth century many astronomical and geographical works containing tables of latitudes were published.[23]

One person who could have produced technical designs for the diptychs was Georg Hartmann (1489–1564). A pastor of St. Sebald's church in Nuremberg, Hartmann was also a mathematician and designer of instruments such as astrolabes, quadrants, and various types of sundials. The existence of several prints signed by Hartmann of diptychs for particular latitudes dated between 1535 and 1553 supports this theory (see p. 12 above). Although none of these designs is found exactly reproduced on any one ivory diptych, they give the impression that Hartmann was in close contact with both engravers and instrument-makers.[24]

Research in the various archives of Nuremberg has yielded valuable biographical information about the six principal families who made ivory diptychs.[25] It has become clear that most of them lived in the parish of St. Lorenz (south of the river Pegnitz), and within this area they were concentrated in the so-called Carthusian quarter, the southwest part of the city (see map on page 38). One of the most striking results of the study is that there were many more compass-makers in the Karner, Ducher, Reinmann, and Miller families than are indicated by surviving ivory diptychs. Unless it is assumed that these instruments are not a representative sample, one conclusion to be drawn is that some members of these families confined themselves to the production of wooden instruments. Another possibility is that while an individual might be described as a compass-maker, he may have worked for another family member or some other individual who actually signed the instruments. A further discovery, which might have been expected, is that the principal compass-making families were related. To give a concrete example: Michael Lesel, active between 1613 and 1629, was the son of the widow of Caspar Karner and made use of a maker's mark normally associated with the Reinmanns.

Some of the sons of compass-makers adopted different trades, and compass-makers themselves often augmented their income by other kinds

23. Gouk (1988), pp. 88–91, 102.

24. Gouk (1988), pp. 91–93; Zinner (1956; reprint, 1979), pp. 357–361.

25. Gouk (1988), pp. 50–62.

of work. Two of the three sons of Hans Ducher II were a peddler and a city wait (that is, a municipal wind musician), while the youngest followed his father's trade. A connection with music also emerges in the case of the compass-maker Albrecht Karner and his three compass-making sons, who were all described in family records as professional violinists. Hans Karner II was at one time a *Hochzeitlader* or professional marriage organizer and later a landlord and victualer, evidently in the same place as his workshop.

As already mentioned, the earliest use of the term *compass-maker* as a profession in the city records dates from the mid-1480s. A group of compass-makers initially attempted to organize their craft in 1510, but the first documented legislation for the compass-makers, as for many other crafts, was formalized by the city council only in 1535. Although compass-making was probably still a *freie Künste* or free craft, these rules imply that the council had by then already allowed the masters to exercise some control over their work. They include details of the masterpiece, the training of apprentices and journeymen, restrictions on production, distribution of materials, and other regulations pertaining to craft practice. During the course of the sixteenth century, new obligations and restrictions were added, so that by 1608, when a revised version of the rules appeared, it was already effectively a closed craft.[26]

The skills necessary to produce compasses were transmitted orally from master to apprentice. Because of a tradition of secrecy, very little is known today about the techniques used in manufacture, apart from what can be gleaned from an examination of the instruments themselves.[27] The regulations of the craft, however, do contain some details about the masterpiece, that is, the work that had to be submitted by a journeyman before being admitted as a master. The basic requirements for this masterpiece remained essentially unchanged from 1535 at least until 1719. Three dozen boxwood compasses with accessories and punches in three different sizes were to be produced by the would-be master; if they were not satisfactory he could not try again for three months.

Several inventories and wills made by compass-makers refer to the tools of their craft. For example, according to the inventory drawn up by his widow Elizabeth in October 1529, Heinz Scheltner left tools for the making of compasses to the value of one guilder. The will of Sebalt Scheltner, dated January 1534, stipulates that his dwelling, his tools, "with all accessories," and eighty guilders should be left to his son Michael, while four ivory teeth (tusks) and some cut bones should be shared between this son and a daughter. The 1544 inventory of the compass-maker Fritz Pilgram and his wife Anna, who both died in 1543, includes the following items: two dials for compasses; about two pounds of cut ivory for compasses; and "what else belongs to the making of compasses." The total value of tools listed was six guilders.[28]

The control of maker's marks or signs was considered an important method of regulating quality.[29] Like the coppersmiths and goldsmiths, each master compass-maker was meant to have one mark with which he signed instruments from his workshop as a means of preventing fraud. It is on the basis of these marks that most surviving instruments can be identified today. The *sworn masters* responsible for enforcing the regulations, as

26. An edited translation of the rules is found in Gouk (1988), pp. 77–81.

27. For the results of the analysis, see Gouk (1988), pp. 69–75.

28. Gouk (1988), p. 67; Nürnberg Stadtarchiv, *Inventarbücher des Stadtgerichts* Rep. B 14, vol. 1, fol. 75v; *Libri Litt.,* vol. 45, fol. 199/199v; *Inventarbücher des Stadtgerichts* Rep. B 14, vol. 3, fol. 54v.

29. Gouk (1988), pp. 66–67, 117; Lockner (1981).

in all the organized crafts, had to check the work of the other masters at various stages of the manufacturing process. If an instrument were of sufficiently high quality the sworn master could stamp on an extra "N" (for Nuremberg) as a sign of approval. This practice, however, seems to have died out by the early seventeenth century.

Investigation into the manufacturing process of the dials has revealed the extent to which the compass-maker could rely on the skills and techniques of other craftsmen in the city. Along with the basic materials of ivory and brass, specialist tools, pigments, and accessories were all immediately available for purchase in Nuremberg. Workmen from all kinds of trades could be readily employed to undertake simple, repetitive tasks. The design of the diptych dial lent itself to serial production of certain basic components in relatively large quantities. Only the final layout of the design and ornament distinguished one dial from another.

A study of the ornamentation of diptych dials reveals the existence of a variety of standardized motifs and patterns common to more than one maker, although individual characteristics can also be identified. The readiest source for the compass-maker's inspiration was the large number of pattern books published in Nuremberg by goldsmiths and engravers from the mid-sixteenth century onwards.[30] Clusters of fruit and vegetables were a form of decoration which originated in Flemish strapwork of the mid-sixteenth century but soon became widespread. Variations of this design, including scrollwork with birds, fruit, and foliage, are found in prints by Bernhard Zan and Hieronymous Bang of Nuremberg, and these motifs first seem to have been used by compass-makers from the late 1620s. It is clear that the compass-makers drew on the same ornamental vocabulary employed by other craftsmen. That this is so is illustrated by weapons made in southern Germany in the late sixteenth century.

Paul Reinmann was the most prominent ivory compass-maker at the beginning of the seventeenth century; his most productive years were between about 1597 and 1608. He used the crown as his mark as his father Hieronymus had done before him. The quality of Paul's instruments,

30. Gouk (1988), pp. 82–83, 94–101; Irmscher (1985).

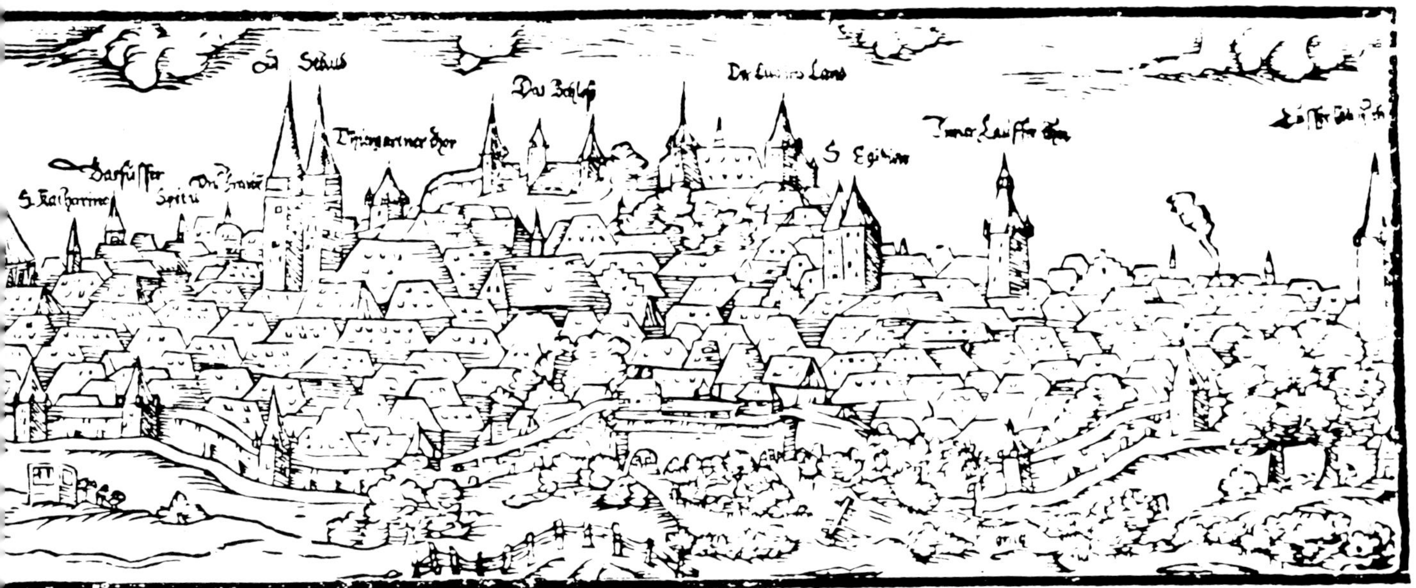

many of which are finely engraved and have gilt brass fittings, reinforce the impression that he was a wealthy craftsman who could afford the best materials and workmanship for his diptychs. It was Reinmann who first extensively adopted the practice of ornamenting the diptychs with fine decorative borders and engraved figures and scenes which were apparently taken from contemporary pattern books, prints, and emblem books.

Reinmann's work represents the apex of the Nuremberg diptych dial. His instruments are notable for their visual attractiveness as well as for their technical accuracy. But the later seventeenth century witnessed a gradual decline in quality. Ornamental standards were maintained, but there was no further improvement to the technical design of sundials. Indeed, the instruments of the later Duchers can be identified by consistent errors in their epact tables. After the middle of the seventeenth century, surviving diptychs are fewer, smaller, cheaper, and altogether less impressive than those that came before. By 1700 the compass-making trade was a mere shadow of what it once had been.

dent. Wiltu gern/ so magstu es mitt teütschen worten schreiben zů der figuren.

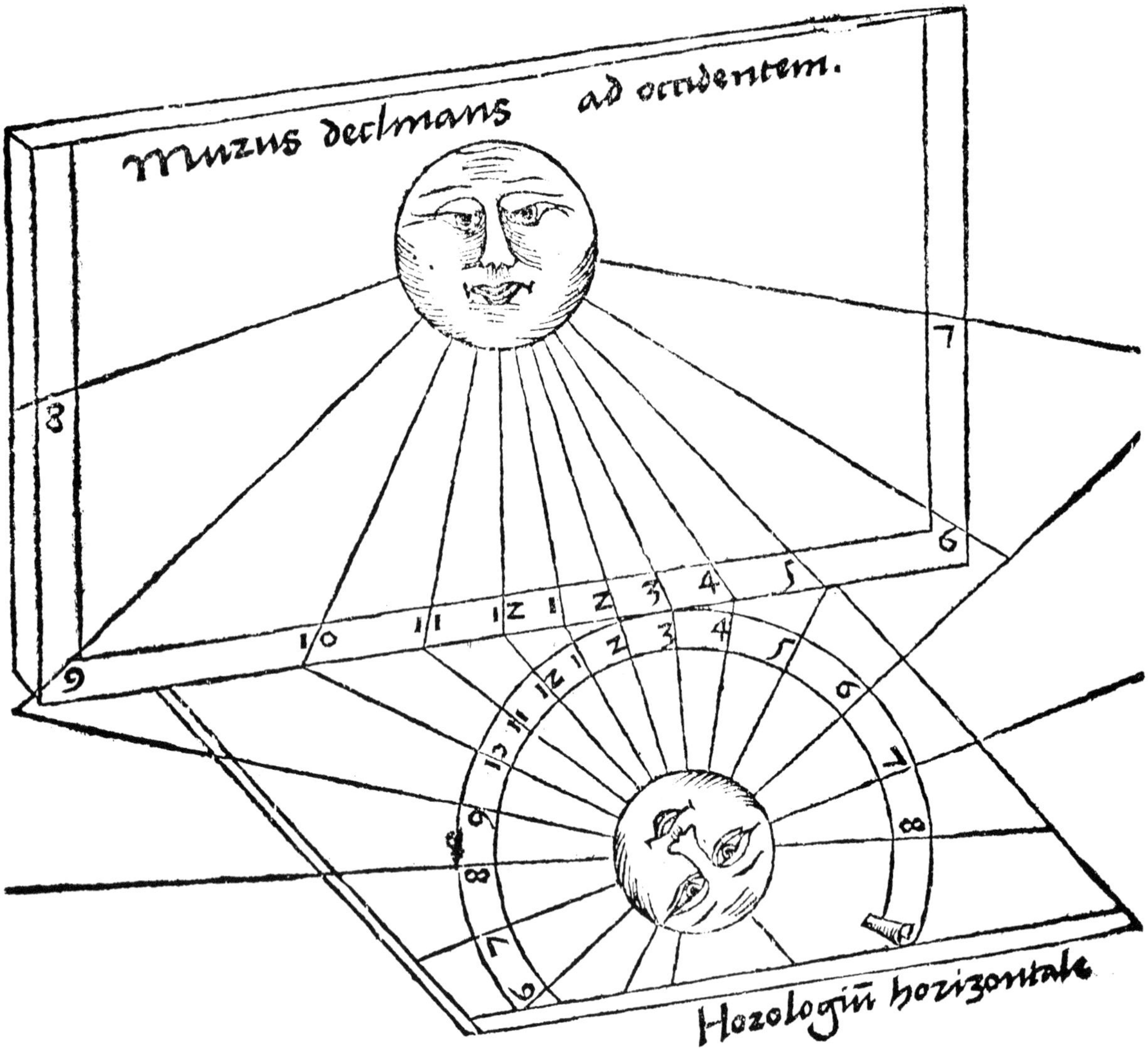

Wie man die Horologia machen soll/ die gerad ghan gegen orient oder occident/ Das xxj capitel.

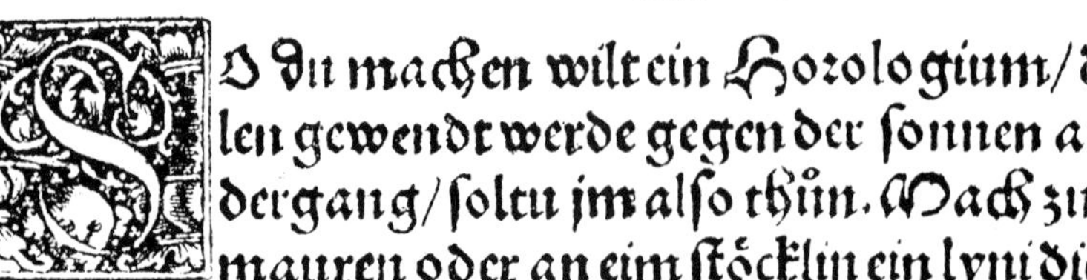

SO du machen wilt ein Horologium/ das on alles fä‌len gewendt werde gegen der sonnen aufgang oder ni‌dergang/ soltu jm also thůn. Mach zum ersten an der mauren oder an eim stöcklin ein lyni die des equinocti‌als höhe hab/ das ist/ die sich von mittnacht gegen dem mittag aufricht/ so vil grad als der equinoctial sich in deinem land erhebt über

The Nuremberg *Kompassmachers*

One can list a few dozen *Kompassmachers* who plied their trade during the mid-sixteenth century, but by the beginning of the seventeenth century there emerged six main families of *Kompassmachers* who passed their skills from one generation to the next: Ducher, Troschel, Miller, Karner, Lesel, and Reinmann. The first five of these families are represented in this catalogue. (All archival information regarding the dates of these makers is derived from Penelope Gouk's *Ivory Sundials of Nuremberg*.) These families lived in close proximity, most of them in the same quarter of Nuremberg (see map on page 38); family connections should not be overlooked. Several diptychs possess what appears to be more than one maker's mark, some from different families. This would seem to indicate either that master craftsmen could send their children to another craftsman's workshop as apprentices, or that a maker's punch marks continued to be used by other craftsmen as decorative marks after the maker's death. Additionally, some diptychs were repaired or modified at a later date, and thus show evidence of two different makers' styles and distinctive marks. Further interconnections between the families arose by intermarriage. The Karner and Lesel families were thus related. Michael Lesel's use of Paul Reinmann's maker's mark (a crown) raises the question of whether he was somehow related to the Reinmann family. Nicolaus Miller's adoption of a similar crown and Joseph Ducher's use of a Troschel-like bird as makers' marks raises the question of continuity within the Nuremberg craft.

The Ducher Family Workshop

Penelope Gouk's investigation at several Nuremberg archives has uncovered at least three different *Kompassmachers* named Hans Ducher. They became masters in 1537, 1557, and 1570. Since the first of these died in 1550, the latter two (referred to as Hans II and Hans III) are responsible for the large output of diptychs dated between 1567 and 1614. Unfortunately, it appears to be impossible to distinguish between these two makers by their punch marks; four slightly different crowned snake punch marks were used between these two over five decades. It is not unreasonable to suppose that punches would break over time and need replacement (as is obvious by comparing the incomplete crown punch marks sometimes used by Paul Reinmann, Michael Lesel, and Nicolaus Miller), and that the replacements would not always be identical to the original punch. While two of these punch marks are encountered much more frequently, it is impossible to assign them respectively to Hans II and Hans III; Thomas Ducher (1590–1645, the son of Hans II) used both marks on instruments dated between 1620 and 1645.

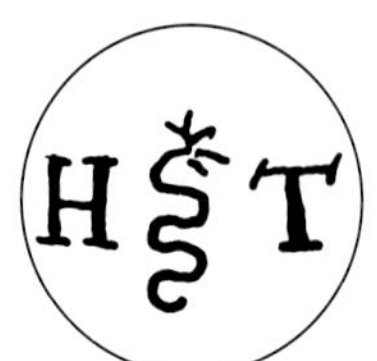

Ducher family maker's marks:

Hans

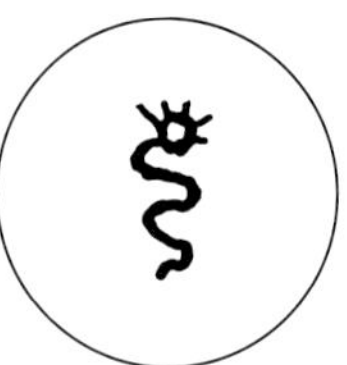

Hans or Thomas

Joseph

One may also try to distinguish between the two Hanses by their signatures. This method of contrast yields a temporal distinction, with perhaps some overlap in the years of productivity. Using diptychs that are both signed and dated, one can establish that the signature "HANS DVCHER" was used in the years 1567–1580, "hans ducher" in 1578–1580, "HANS TVCHER" in 1582–1589, and the initials "HD" in 1595–1600. (See Appendix II.) None of the diptychs both dated and signed "HD" were "typical" rectangular diptychs, but rather oval, round, or polyhedral; there are even triptychs (one example is in the Adler Planetarium, inv. no. DPW-29)! This means of distinguishing between the dials by the signature aids in assigning dates, but it does not solve the problem of which Hans is which. All of the "Hans Ducher" dials in this catalogue (cat. nos. 1–4) fall into the later temporal category; they are also all oval or egg-shaped and appear to have been produced by the same Hans, whichever one he may have been.

Most of Hans Ducher's dials produced in the 1590s were adjustable for use in northern Italy. Some were made solely for use in Italy, as demonstrated by the phrase "TEVTSCH VND WELSCH VHR IN VENETIA" (German and foreign hours in Venice, cat. no. 2) and "AD ELEVATIO POLI 43 GRAD" (at a latitude of 43 degrees, cat. no. 3), as well as by the inclusion of Italian hour scales on all of his dials. These dials were presumably either manufactured for export to Venice, one of Nuremberg's major trading partners, or intended for purchase by German merchants traveling south of the Alps.

It is possible, however, to distinguish between unsigned instruments by the Hans Duchers and those of Thomas on ornamental and technical grounds. The decoration of the early Ducher instruments (before about 1610) consists almost entirely of Latin and German inscriptions in Roman and Gothic script, with the coloring limited to black and red. These inscriptions often include instructions on how to use the instrument. One of the most common inscriptions is the rhyming German verse, "ICH

CAMPAST KAN NICHT RECHT WEISEN NAHET BEY EISSEN" (see cat. no. 3), which roughly translates as "I, the compass, cannot indicate correctly when placed near iron." The inclusion of the Latin term LIBELLO on one of Hans's diptychs (see cat. no. 1) may mean that the maker provided a small printed booklet with instructions on how to use the dial. While there is evidence for such booklets in the seventeenth and eighteenth centuries (accompanying Bloud-type ivory dials made in Dieppe and brass Augsburg-type horizontal dials), this inscription from the late sixteenth century appears to be the earliest reference to such booklets. Also, the dials made by the Hans Duchers sometimes include materials other than ivory, such as wood or gilded brass.

In contrast, many of the instruments attributable to Thomas Ducher are elaborately decorated with scrollwork of fruit, flowers, and leaves. Thomas is also well-known for his diptychs resembling small books with elaborate clasps, hinged on the side rather than at the top (cat. nos. 6 and 7). His colors include green, blue, and a rusty orange, as well as black and red. Thomas's dials also sometimes incorporate characteristic designs, such as the triple-ringed "globe" figure and gilt spandrels. His epact tables are likewise unique: he always began them with the same sequence (17, 28, 9… and 7, 18, 29…) and mislabeled the Gregorian and Julian epacts.

Joseph Ducher (1614–1644), Thomas's son, produced many colorful diptychs in the style of Lienhart/Leonhart Miller (compare cat. no. 9 with nos. 18 and 20). It is curious that he adopted the figure of a bird flying to the left as his maker's mark, rather than using the crowned snake of his father and grandfather. A similar mark was associated with the maker Hans Troschel the Elder. Joseph's instruments are ornamented with scrollwork, usually with characteristic hook-shaped leaves, and are colored with green, orange, black, and red.

1 Oval Ivory Diptych

signed "HT" [Hans Ducher] and
maker's mark
c. 1585
Nuremberg, Germany
5.6 (w) x 7.0 (l) x 1.3 (h) cm
Formerly Ernst Collection No. 30
Inventory No. 7830

⚙ Black, red, and blue coloring. Brass
wire clasps on I and II, hanging loop and
four bun feet on II. Compass needle and
glass intact. Plate II is made from two slices
of ivory glued together. The exterior of II
is cracked in several places, and I is split
lengthwise.

Ia Standard German lunar volvelle with
stamped brass rotating disc (2.3 cm) with
index. Scales numbered 1–12 twice (inner
and middle) and 1–29 (outer). Lunar phase
indicated in red and blue under disc. Man-
in-the-moon motif on volvelle. Outer
circle labeled "PLENIP. VNIVM," "PRIMA
QVARTA," "NOVIL. VNIVM," and "VLTIMA
QVARTA" [full moon, first quarter, new
moon, last quarter]. Two brass plates in the
shape of lunar crescents hold ivory together.

Ib Holes for string gnomon at 42°, 45°,
48°, 51°, and 54°. "VTI HOC COMPASS
LICET UBIQVE VT DE INSTRVCTIONE EIVS EX
LIBELLO PERCIPIAS" [It is possible to use
this compass anywhere, so observe the
instruction from the booklet] and "IMPONE
FILVM FORAMINI QVOD LOCI TVI POLVM
INDICAT ET HORA HABEBITVR IN CIRCVLO
QVI EODEM NUMERO POLARI INSCRIBITVR"
[Place the string into the hole which
corresponds to your latitude and the time
will be indicated in the circle which is
inscribed with the same latitude].

IIa Horizontal dial with five scales, middle
and outer scales labeled 4–12–8 and inner
scale labeled 5–12–7, marked for latitudes
54° (outer), 51°, 48°, 45°, and 42° (inner).
Compass needle and glass intact. Compass
points labeled S, ORIENS, M, and OCCIDENS
and extraneous gilt indicator in compass
bowl (added later?). Magnetic declination
6° east of north. Initials "HT" with
crowned snake maker's mark in between.

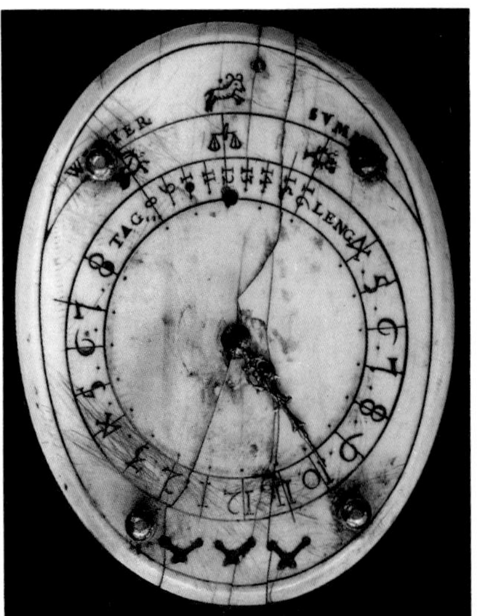

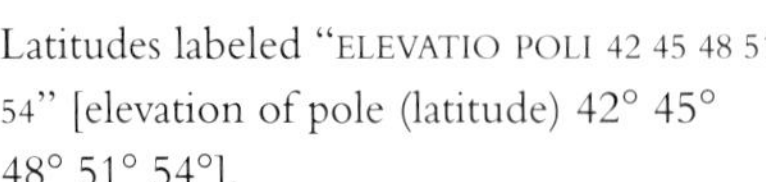

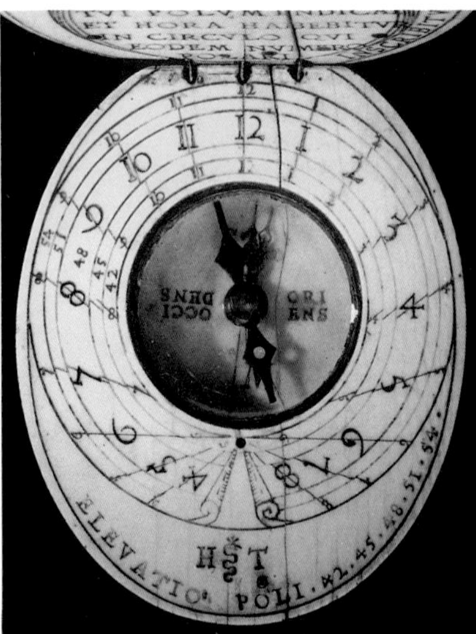

Ia	Ib
IIb	IIa

Latitudes labeled "ELEVATIO POLI 42 45 48 51
54" [elevation of pole (latitude) 42° 45°
48° 51° 54°].

IIb Scale reading 4–12–8, with gilt indicator (attached to the one in the compass
bowl). This indicator is apparently a useless
replacement for the original brass volvelle

used to convert between large and small
hours. Scale for length of the day reading
8–16 labeled "TAG LENG" and "WINTER"
and "SVMMER," with four zodiacal
symbols. Image inverted for legibility.

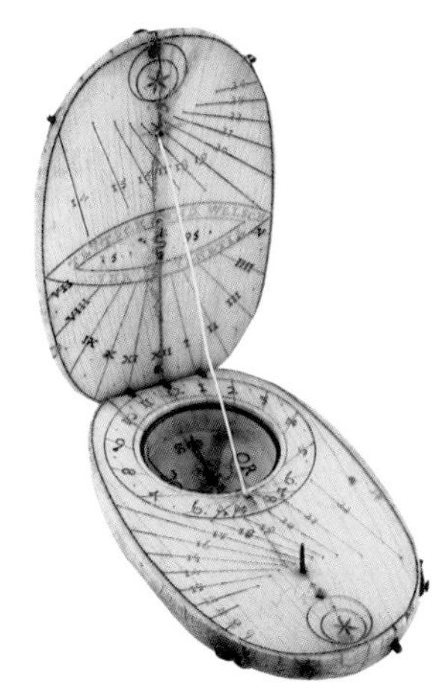

2 Oval Ivory Diptych

signed "HD" [Hans Ducher] and
maker's mark
dated 1595
Nuremberg, Germany
5.4 (w) x 8.8 (l) x 1.2 (h) cm
Formerly Drecker Collection No. 116,
then David P. Wheatland Collection
Inventory No. 7574

◉ Black and red coloring. Tablet I is split
lengthwise and has been glued together.
Brass wire clasps on I and II, bun foot on I.
The two tablets of ivory have become
separated. Small hole in II where compass
pin was once attached. Two dark splotches
on II. Compass needle and glass missing,
brass ring intact.

Ia Undecorated. Remnants of glue and
paper remain along central seam where the
two halves were glued back together.

Ib Pin gnomon dial (gnomon missing)
indicating Italian hours, labeled 14–24.
Inscribed "TEVTSCH VND WELSCH VHR IN
VENETIA" [German and foreign hours in
Venice]. Maker's mark (crowned snake)
amidst date "1595." Vertical dial, labeled
VII–XII–V in black numerals, with dots
marking the half-hours. Six-pointed star
and lunar crescent motif at top.

IIa Horizontal dial with a single scale for
approximate latitude 56°, labeled 4–12–8
in black numerals, with dots marking the
half-hours. Compass points labeled S, OR,
M, and OC. Magnetic declination indicated
at 7° east of north by a fleur-de-lys. Pin
gnomon dial for Italian hours labeled 9–23
in red numerals. Six-pointed star and lunar
crescent motif.

IIb Undecorated. Initials "HD" with
maker's mark (crowned snake) between
initials appear at center. Numbered 216 in
black ink. Image inverted for legibility.

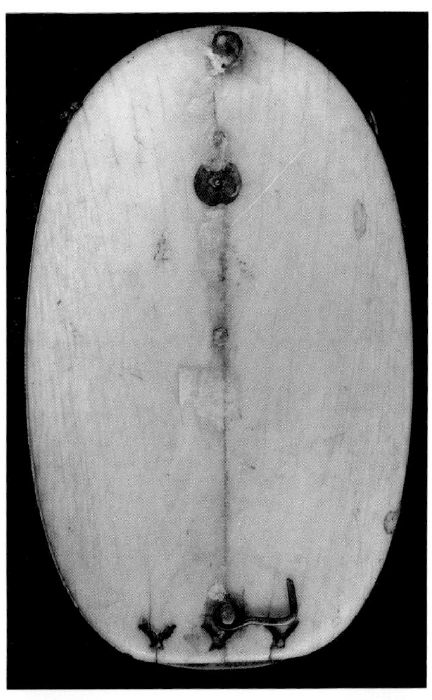

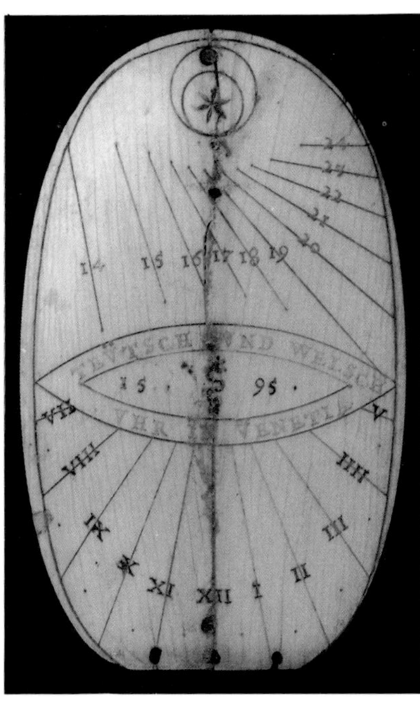

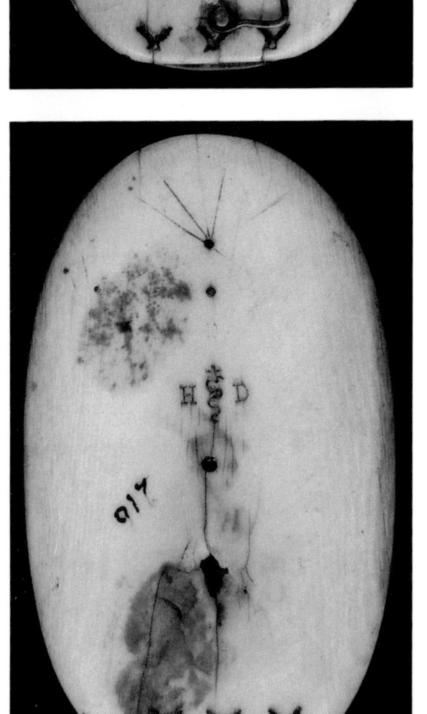

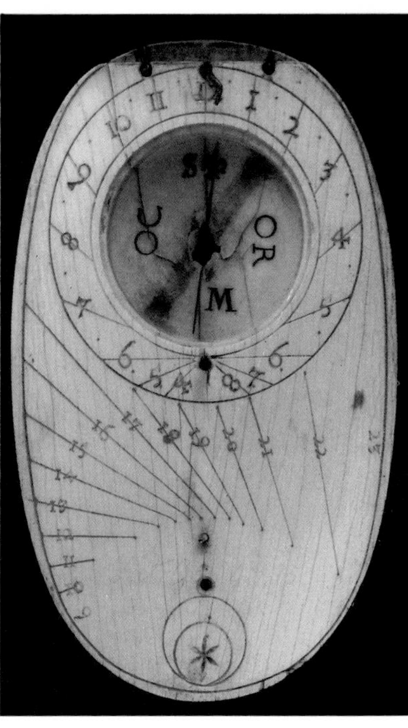

3 Oval Ivory and Wood Diptych

signed with maker's mark [Hans Ducher]
c. 1595
Nuremberg, Germany
4.9 (w) x 6.5 (l) x 1.5 (h) cm
Formerly Drecker Collection No. 118,
then David P. Wheatland Collection
Inventory No. 7575

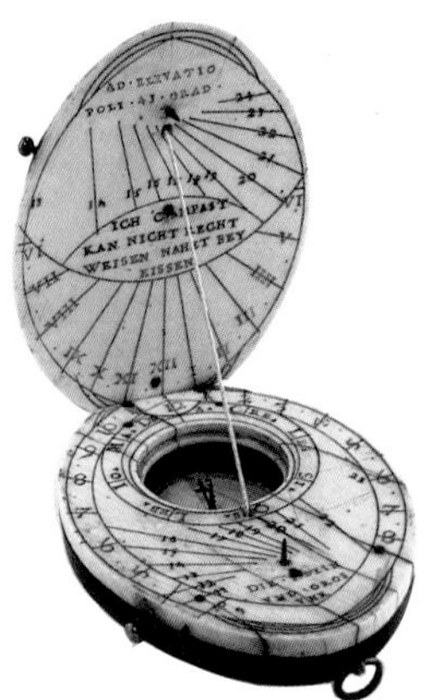

⬤ Black and red coloring. Tablet I is a
thin slice of ivory, II is a sandwich of two
slices of ivory with a rosewood core. Brass
wire clasps on I and II, hanging loop on II.
One of the two wire hinges is broken.
Compass needle and glass missing.

Ia Undecorated.

Ib Pin gnomon dial indicating Italian
hours (labeled 12–24) reads: "AD ELEVATIO
POLI 43 GRAD" [elevation of pole (latitude)
43°]. "ICH CAMPAST KAN NICHT RECHT
WEISEN NAHET BEY EISSEN" [I, the compass
sundial, cannot indicate correctly when
placed near iron]. Vertical dial, labeled
VI–XII–VI in red numerals, with dots
marking the half-hours.

IIa Horizontal dial with a single scale for
approximate latitude 43°, labeled 5–12–8
in red numerals, with dots marking the
half-hours. Directions of the 8 winds
abbreviated in Italian around the compass
bowl: "DRA GRE LE SI OS LEB IO MA."
Crowned snake maker's mark in compass
bowl. Magnetic declination indicated at 8°
(solid line) and 16° (dotted line) east of
north. Pin gnomon dial for Italian hours
labeled 9–23. Inscribed "DIE KLEIN VND
GROS VHR" [the small and large hours].

IIb Undecorated. Maker's mark (crowned
snake) appears once along with the
Nuremberg "N" stamp (very faint).
Numbered 93 twice, in red ink and in
black on a small modern paper label.

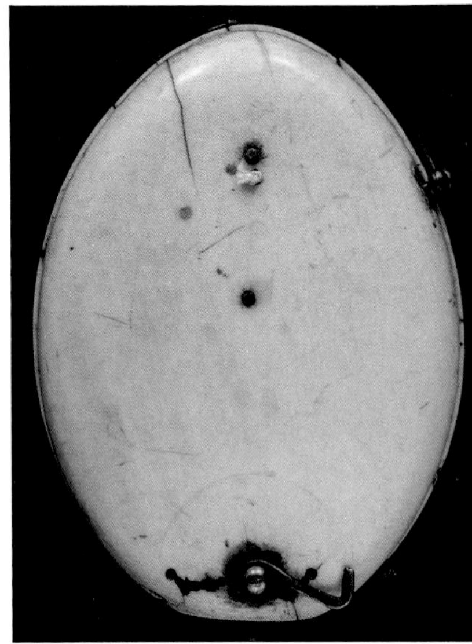

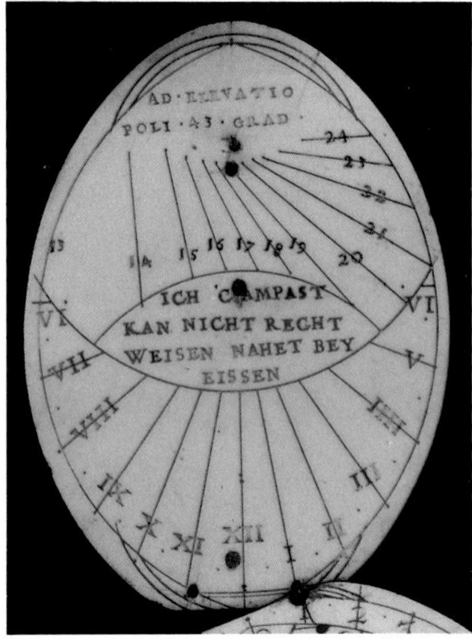

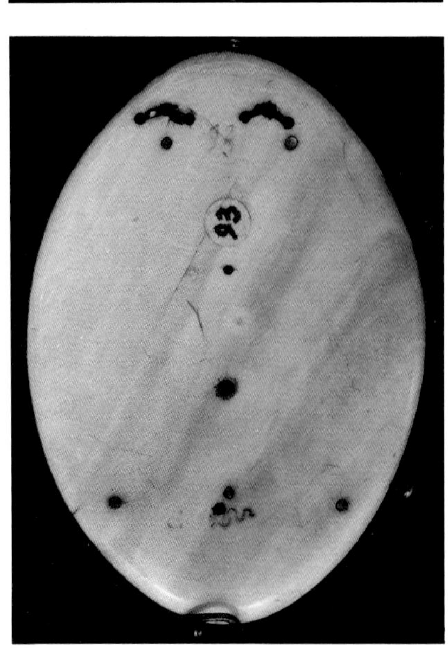

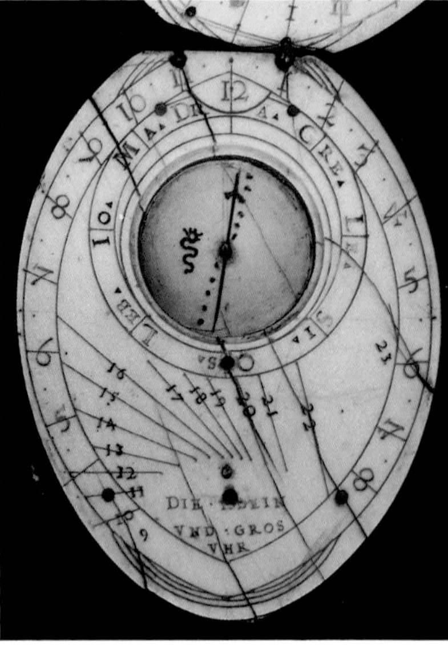

4 Oval Ivory Diptych

signed with maker's mark [Hans Ducher]
dated 1611
Nuremberg, Germany
7.4 (w) x 8.5 (l) x 1.2 (h) cm
Formerly David P. Wheatland Collection
(purchased in France, 1952)
Inventory No. 7577

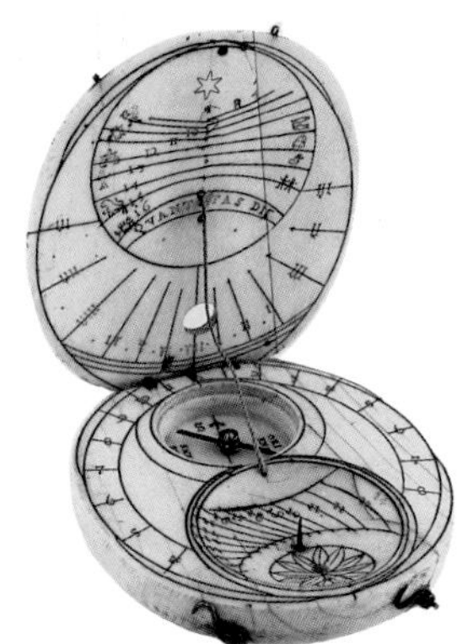

⚙ Black, red, and blue coloring. Tablet I has been spliced together. Compass viewing-hole in I, no slot for wind vane in II. Brass wire clasps on I and II. Compass needle and glass are replacements.

Ia Sixteen-point wind rose with scale reading 2 to 32 in twos, 8 directions labeled in German and 4 winds labeled in Italian (OSTRO, PENETE, TRAMONTARIA, and LEVANTE). Hand-shaped brass indicator. Fleur-de-lys stamp near the hinge indicates north.

Ib Pin gnomon dial (gnomon missing) with scale reading 8–16 in red numerals labeled "QVANTITAS DIE" [length of the day] and marked with zodiacal symbols. Vertical dial, labeled VI–XII–VI in Gothic-style numerals (in red, except last two are black), with dots marking the half-hours. Six-pointed star above the pin gnomon dial.

IIa Horizontal dial with a single scale for approximate latitude 49°, labeled 4–12–8 in red numerals. Directions labeled S, ORIENS, M, and OCCIDENS. Magnetic declination 7° east of north. Scaphe dial with rosette is 4 cm in diameter and indicates Italian hours (labeled 8–24 in red numerals).

IIb Standard German lunar volvelle labeled "COMMPASSVS NOCTVRNVS" [night compass sundial], brass disc missing. Scales labeled 1–12 twice (inner) and 1–29 (outer). Relative phase of moon for use with brass disc indicated by red and blue engraving. Circular motif. Maker's mark (crowned snake) inserted in the date "1611." Image inverted for legibility.

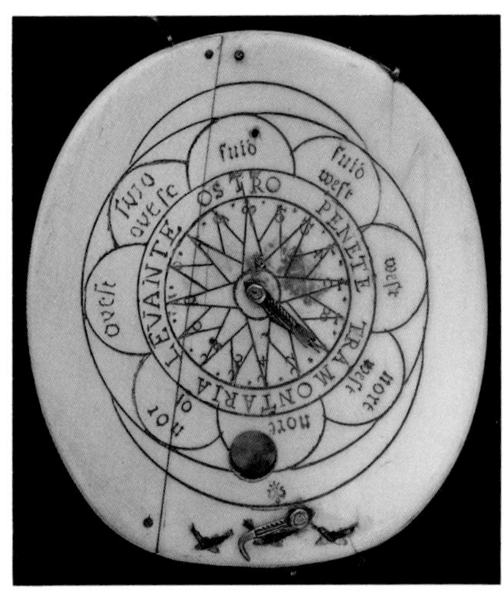

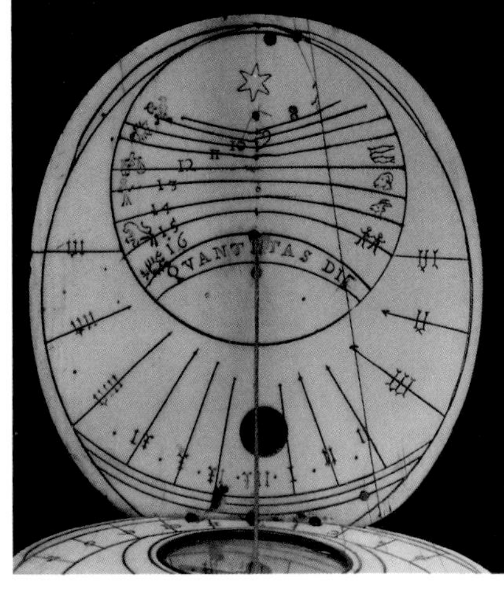

Ia | Ib
IIb | IIa

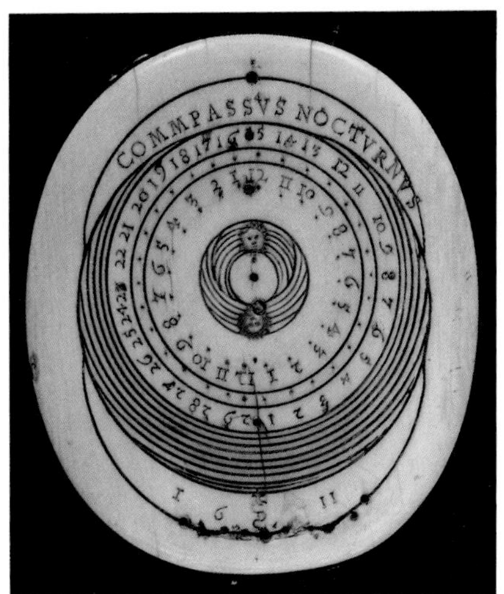

5 Rectangular Ivory Diptych

signed "TD" [Thomas Ducher] and
maker's mark
before 1645
Nuremberg, Germany
6.4 (w) x 8.0 (l) x 1.3 (h) cm
Formerly Drecker Collection No. 121,
then David P. Wheatland Collection
Inventory No. 7576

⚙ Faded black and red coloring. Outer faces are yellowed, especially II. Corner of I is reconstructed. Compass viewing-hole in I, no slot for wind vane in II. Brass wire clasps on I and II, two bun feet on IIb. Compass needle and glass missing, brass ring intact.

Ia Wind rose with 16 points, scale numbered 1 to 32 (east = 32), 7 directions labeled in German, cardinal directions labeled S, ORIENS, M, and OCCIDENS. Indicator missing. Inscriptions "WEG WEISER" [path finder] and "CAMPASSVS" [compass/ portable sundial]. Small decorative stamps in corners. Edges bordered by a triple- and single-lined "frame."

Ib Pin gnomon dial labeled 8–16 in red numerals, inscribed "TAG LENG" [length of the day], and marked with zodiacal symbols. Vertical dial labeled VII–XII–V in Gothic-style red numerals, with lines and dots marking the half- and quarter-hours, respectively. Decoration includes six-pointed star and a crescent moon at center. Triple- and quadruple-lined border.

IIa Horizontal dial with a single scale for approximate latitude 47.5°, labeled 4–12–8 in black numerals, with lines and dots marking the half- and quarter-hours, respectively. Compass bowl contains maker's mark (crowned snake) twice. Cardinal points labeled S, ORIENS, M, and OCCIDENS. Magnetic declination indicated for 7° east (with a fleur-de-lys) and 18° west (with a line) of north. Pin gnomon dial for Italian hours labeled 9–22 in red numerals inscribed "WELSCH VR" [foreign hours]. Small decorations in corners. Double- and triple-lined border.

IIb Standard German lunar volvelle labeled "NACHT VHR" [night clock]. Brass disc

missing. Scales labeled 1–12 twice (inner) and 0–29 (outer). Gregorian and Julian epacts (beginning with 17 and 7, respectively, for the year 1621 or 1640) tabulated circularly around volvelle and labeled "EPAGTA GREGORI" and "EPAGTA ILIANA" [Gregorian and Julian epacts]. Note that the names of the epacts are reversed. Initials "TD" with maker's mark

between initials appears within outermost oval. Small decorations with stars in corners. Numbered 201 or 207 in red near center. Double- and single-lined "frame" border.

6 Rectangular Ivory Diptych Book

signed with maker's mark [Thomas Ducher]
before 1645
Nuremberg, Germany
6.0 (w) x 8.2 (l) x 1.4 (h) cm
Formerly Drecker Collection No. 119,
then David P. Wheatland Collection
Inventory No. 7581

⊙ Faded black, red, and green coloring. Hinged lengthwise. Tablet IIb is badly discolored (burned). Compass viewing-hole in I, no slot for wind vane in II. Brass wire clasps on I and II, four bun feet on Ia and IIb. Compass needle intact and glass missing, brass ring intact. Edges of each tablet bordered by a "frame" that includes alternating circles and stars.

Ia Wind vane with scale of 1 to 32 (east = 32), 8 directions labeled in German, cardinal directions labeled S, OR, M, and OC. Brass dagger indicator. Foliate decoration.

Ib Circular vertical dial, labeled VII–XII–V in Gothic-style black numerals, with lines and dots marking the half- and quarter-hours, respectively. Decoration includes six-pointed star surrounded by 8 small star stamps, a crescent moon, and engraved foliage.

IIa Horizontal dial with a single scale for approximate latitude 45°, labeled 4–12–8 in black numerals, with lines and dots marking the half- and quarter-hours, respectively. Compass bowl contains maker's mark (crowned snake) twice. Cardinal directions labeled S, ORIENS, M,

and OCCIDENS. Magnetic declination indicated at 4° east (fleur-de-lys) and 20° west (faintly scratched line) of north. Foliate decoration.

IIb Standard German lunar volvelle, brass disc missing. Scales labeled 1–12 twice (inner) and 0–29 (outer). Engravings of pomegranates and foliage.

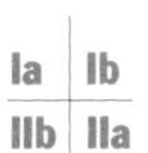

Ia	Ib
IIb	IIa

7 Rectangular Ivory Diptych Book

unsigned [Thomas Ducher]
before 1645
Nuremberg, Germany
7.8 (w) x 12.5 (l) x 1.3 (h) cm
Formerly Drecker Collection No. 120,
then David. P. Wheatland Collection
Inventory No. 7579

☯ Faded black, red, and green coloring. Hinged lengthwise. Corner of II chipped. Compass viewing-hole in I. No slot for wind vane. Brass hook clasps on I, four flattened bun feet on IIb. Compass needle, glass, and brass ring intact. Edges of each tablet bordered by a double- and single-lined "frame."

Ia Wind rose with 16 points, numbered 1–32 in German (east = 32), with arm-shaped indicator. Sixteen directions labeled in German, cardinal points labeled twice in Latin as S, ORIENS and OR, M, and OCCIDENS and OC. Engraving of an Ottoman Turk and an American Indian (?) facing each other on either side of the wind rose among foliage.

Ib Vertical dial labeled VI–XII–VI in Gothic-style black numerals, with lines and dots marking the half- and quarter-hours, respectively. Holes for string gnomon at 40°, 42°, 45°, 48°, and 54° latitude, with intermediate holes. Colorful engraving of vegetation and foliage.

IIa Horizontal dial with three scales labeled 4–12–8 in black (inner), IIII–XII–VIII in red (middle), and 4–12–8 in black (outer), with lines and dots marking the half- and quarter-hours, and marked 42°, 48°, and 54°, respectively. Compass bowl has no apparent maker's mark, but two small six-pointed star stamps. Cardinal directions labeled S, ORIENS, M, and OCCIDENS. Magnetic declination indicated at 2° east (fleur-de-lys) and 15° west (faintly scratched line) of north. Two pin gnomon dials at bottom, on left labeled "H.B." [horae babyloniae] and 2–15 in black numerals for Babylonian hours, on right labeled "H.I." [horae italianae] and 9–22 in red numerals for Italian hours. Foliate decoration.

IIb Standard German lunar volvelle with a stamped brass rotating disc (2.4 cm) with index. Scales labeled 1–12 twice (inner and middle) and 1–29 (outer). Man-in-the-moon motif on brass volvelle. Depictions of "NEI MON," "ERST VIRTEL," "VOL MON," and "LETZT VIRTEL" [new moon, first quarter, full moon, and last quarter] on ivory. Engraving of a European man and woman amidst foliage. Image inverted for legibility.

"

8 Rectangular Ivory Diptych

signed "TD" [Thomas Ducher] and
maker's mark
before 1645
Nuremberg, Germany
5.8 (w) x 10.3 (l) x 1.4 (h) cm
Formerly David P. Wheatland Collection
(purchased from Sotheby's 13 March 1967)
Inventory No. 7573

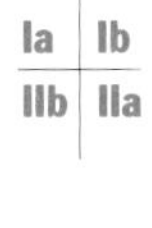

⊛ Black, red, and green coloring, gilt
used around borders. Tablet I is split
sideways twice, II once. Both plates were
spliced back together, apparently by the
maker. Brass plate on right side of II with
initials "TD" and maker's mark (crowned
snake) between initials. Compass viewing-
hole in I, slot for wind vane has been filled
in. Brass wire clasp on II and four bun feet
on IIb. Compass needle intact, glass is a
replacement.

Ia Wind rose with 16 points and scale
numbered 1–32 (east = 32), 8 directions
labeled in German, cardinal directions
labeled twice as S, ORIENS and OR, M, and
OCCIDENS and OC. Indicator broken off.
Engraving of the heads of an Oriental man
and a Turkish man (?) amidst colorful
foliate decoration.

Ib Vertical dial labeled VI–XII–VI in Gothic-
style black numerals, with lines and dots
marking the half- and quarter-hours,
respectively. Colorful foliate decoration
surrounds an elaborate sun face.

IIa Horizontal dial with a single scale for
approximate latitude 48.5°, labeled 4–12–8
in black numerals, with lines and dots
marking the half- and quarter-hours,
respectively. Compass bowl contains
maker's mark (crowned snake) twice.
North and south labeled S and M, respectively. Magnetic declination indicated at 1°
west of north by a fleur-de-lys. Pin
gnomon dial with sun face indicates Italian
hours (labeled 10–22 in red numerals) and
Babylonian hours (labeled 2–14 in black
numerals). Colorful foliate decoration.

IIb Standard German lunar volvelle, brass
disc missing. Scales labeled 1–12 twice in
red numerals (inner) and 1–29 in black
numerals (outer). "EPAGTA GREGORI" and

"EPAGTA IVLIANA" [Gregorian and Julian
epacts] (beginning with 17 and 7,
respectively, for the year 1621 or 1640)
tabulated circularly around volvelle. Note
that the names of the epacts are reversed.
Colorful foliate decoration.

9 Rectangular Ivory Diptych

signed with maker's mark [Joseph Ducher]
c. 1642
Nuremberg, Germany
7.7 (w) x 12.5 (l) x 1.3 (h) cm
Formerly Department of Printing and
Graphic Arts, Houghton Library, Harvard
University (on loan); gift of Philip Hofer
Inventory No. 7899

⊙ Black, red, green, and brown coloring.
Compass viewing-hole in I, slot with brass
clasp for wind vane in side of II. Brass wire
clasps on I and II (one broken), four bun
feet on each Ia and IIb, brass hanging loop
on II. Compass needle, glass, and brass ring
intact.

Ia Wind rose with 32 numbered points, 16
directions labeled in German as *nort, ost,
sud,* and *west,* and 8 winds labeled in
Italian. Brass hand-shaped indicator. Wind
vane missing. Solar motif at center of wind
rose. Four-winds decoration in corners.
Colorful foliate decoration.

Ib Pin gnomon dial calibrated 8–16 (left)
and 16–8 (right) for the length of the day
and night, respectively, marked with
zodiacal symbols and hour lines for Italian
hours (13–23) and Babylonian hours (1–10),
all in black numerals. Engraving of a walled
village and a castle in brown on a green
foreground. Gilded smiling sun motif with
foliate decoration. Holes for string
gnomon at 42°, 48°, and 54°. Latitude list
of 24 cities (42°–54° latitude).

IIa Horizontal dial with three scales
labeled 6–12–6 in black numerals (inner),
IIII–XII–VIII in red stylized Gothic numerals
(middle), and 4–12–8 in black numerals
(outer), marked for latitudes 42°, 48°, and
54°, respectively. Maker's mark (bird flying
to the left) appears twice in compass bowl,
along with four-winds motif. Magnetic
declination 16° west of north, geographic
north indicated by a fleur-de-lys. Two
identical 3.5 cm scaphe dials with sun-face
motifs and pin gnomons indicating Italian
hours, labeled 10–23 in black numerals.
Colorful foliate decoration.

IIb Standard German lunar volvelle with a
stamped brass rotating disc (2.7 cm) with

index, a man-in-the-moon motif, and a
scalloped edge. Scales labeled 1–12 twice
(inner and middle) and 1–29 (outer).
Depictions of "nei man," "erst virtel," "fol
man," and "letzt virtel" [new moon, first
quarter, full moon, and last quarter].
Circular epact table for Gregorian and
Julian epacts labeled "EPAGTA GREGORI"
and "EPAGTA IVLLIANA" and beginning
with 17 and 7, respectively, for the year
1640. Note that the names of the epacts
are reversed. Four-winds decoration in
corners. Foliate decoration.

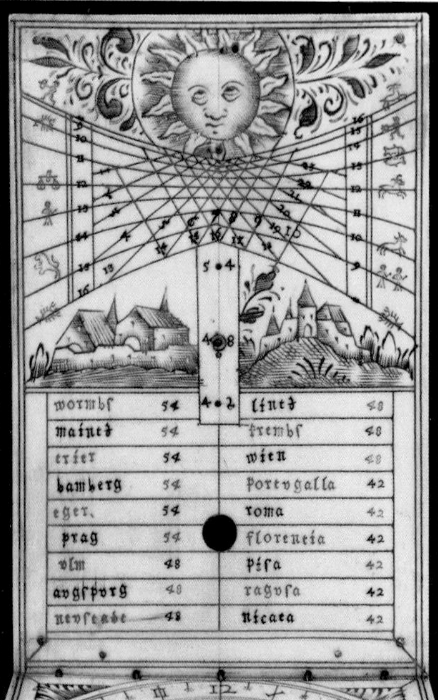

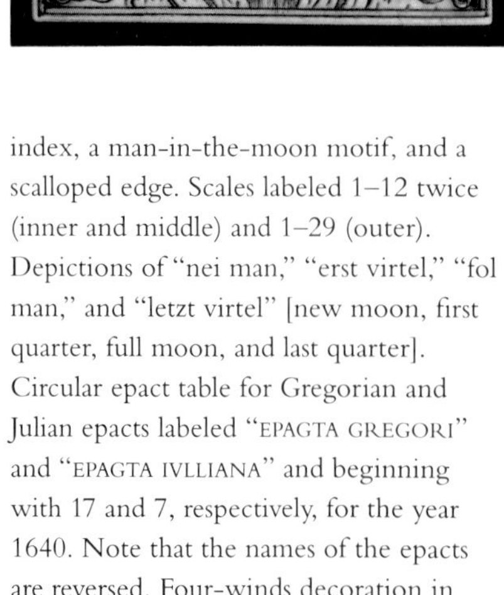

The Troschel Family Workshop

There were two Hanses in the Troschel family, who were father and son and who produced dated instruments in the period 1580–1631 (see Chandler and Vincent, 1969, pp. 211–212). Troschel instruments are usually colored in black and red and frequently include mottos or other inscriptions. Both of the Troschels were skilled engravers (see especially cat. no. 12). Troschel dials exhibit a wide range of technical skill, from complex dials with little ornamentation (cat. no. 11) to extremely simple dials with elaborate engraving (cat. no. 15). The Troschels were one of the few Nuremberg *Kompassmachers* of the seventeenth century to incorporate equinoctial dials in their diptychs (cat. nos. 12 and 13). The Troschels also manufactured many diptychs for use at latitudes corresponding to northern Italy.

Hans Troschel the Elder (1549–1612) used the maker's mark of a thrush on a twig, thus making a word play on his name (*Drossel* means "thrush" in German, which in the Franconian dialect would be pronounced the same as *Troschel*). There are at least two different versions of his maker's mark, one with the bird facing right, the other facing left. Also, cat. no. 10 bears two different versions of this thrush-on-a-twig mark, one stamped carefully, the other scratched rather crudely. While Hans the Younger (1599?–1634) typically used a six-pointed star as his maker's mark, he might also have used his father's second thrush mark on some of his dials as well.

Hans Troschel the Elder. Early-seventeenth-century engraving. (Private collection.)

Troschel family
maker's marks:

Hans the Elder

Hans the Younger

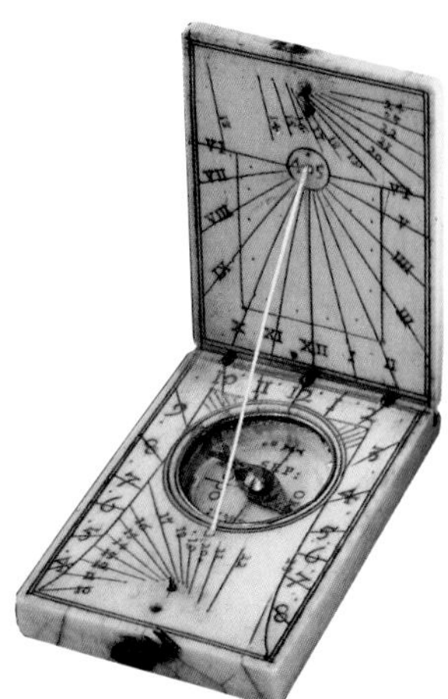

10 Rectangular Ivory Diptych

signed with maker's mark
[Hans Troschel the Elder]
c. 1590
Nuremberg, Germany
4.1 (w) x 6.0 (l) x 1.2 (h) cm
Formerly Drecker Collection No. 122,
then David P. Wheatland Collection
Inventory No. 7533

◉ Black and red coloring. Black coloring has a faded or brownish tint. Brass horse-head clasps on I and II. Compass needle, glass, and silvered ring intact. Numbered 208 along hinge in black ink.

Ia Undecorated. Several scratched letters are barely recognizable (in Italian?) and the date 1628.

Ib Pin gnomon dial labeled 13–24 in black numerals for Italian hours. Vertical dial labeled VI–XII–VI in black numerals, with dots marking the half-hours. Hole for string gnomon is labeled 45 in red.

IIa Horizontal dial for 45° latitude, labeled 4–12–8 in black numerals, with dots marking the half-hours. Cardinal directions in compass bowl labeled SEP:, ORI:, MERI:, and OCCI:. Magnetic declination 8° east of north. Pin gnomon dial for Italian hours labeled 10–23 in black numerals.

IIb Undecorated. The maker's mark, a bird facing right on a twig, appears twice, one stamped in the usual manner, the other a cruder version scratched onto the ivory, probably by some imitator. A very faint "N" punch mark (the mark of a Nuremberg dial) to the left of and slightly below the upper bird. Several other scratched letters are discernible. Image inverted for legibility.

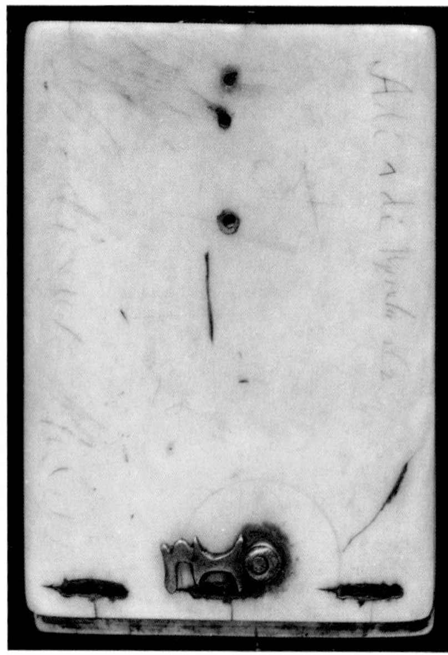

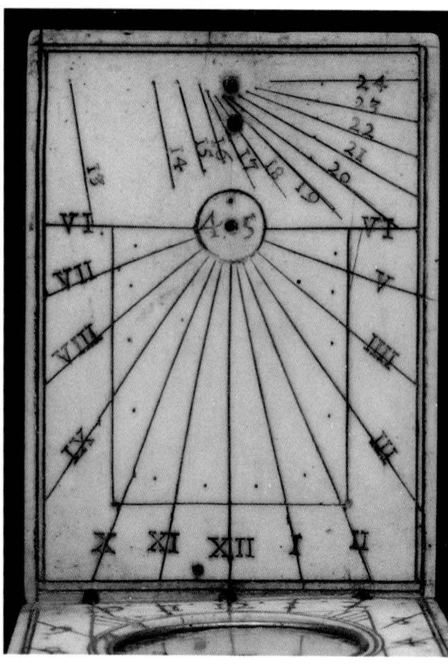

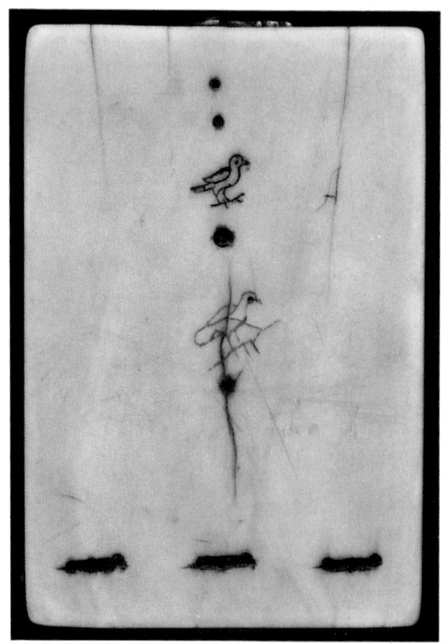

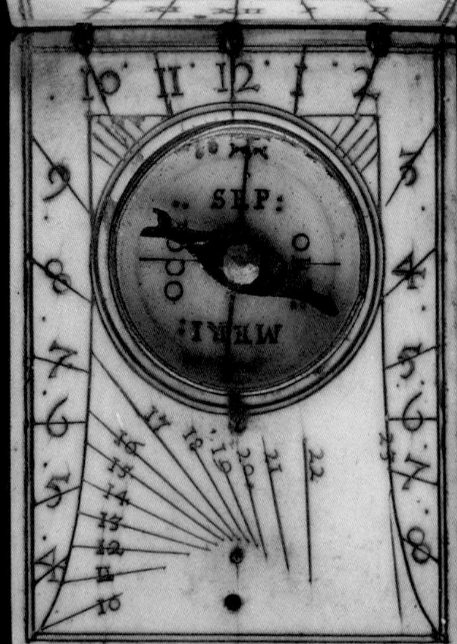

la | lb
llb | lla

11 Rectangular Ivory Diptych

signed "HANS TROSCHEL" [the Elder] and
maker's mark
dated 1600
Nuremberg, Germany
6.0 (w) x 9.1 (l) x 1.2 (h) cm
Formerly Harold Gillingham Collection
No. 82, then David P. Wheatland
Collection
Inventory No. 7537

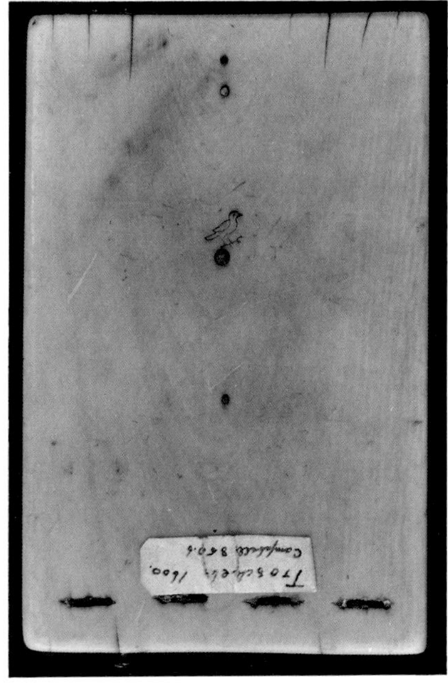

⚫ Black and red coloring. Brass scythe clasps on I and II. Compass needle and glass missing.

Ia Standard German lunar volvelle with a stamped brass rotating disc (2.6 cm) with index and man-in-the-moon motif. Scales labeled 1–12 twice (inner and middle) and 1–29 (outer). Epact table labeled "EPACTA IVLIA" and "EPACTA GREGORI" [Julian and Gregorian epacts] in circular form around volvelle, beginning with 3 and 23, respectively, for the year 1598. Inscribed "AETAS LVNAE ET HORAE NOCTIS" [age of the moon and hour of the night].

Ib Pin gnomon dial labeled "tag leng" [length of day], with hour lines for common hours (labeled VIII–XII–IIII in red) and the length of the day (labeled 8–16 twice in black), marked with zodiacal symbols. Vertical dial labeled 6–12–6 in red numerals, with lines and dots marking the half- and quarter-hours, respectively. Large sun-face motif and foliate decoration. Signed, "HANS TRÖSCHEL NORNBERGAE FECIT ANNO DOMINI 1600" [made by Hans Troschel of Nuremberg in the year of our Lord 1600].

IIa Horizontal dial with a single scale for approximate latitude 47.5°, labeled 4–12–8, with lines and dots marking the half- and quarter-hours, respectively. Compass bowl pasted over with printed paper wind rose insert in French. Magnetic declination (as seen beneath the paper compass bowl insert with a bright light) 8° east of north. Four compass directions in Latin (SEPTENTRIO, ORIENS, MERIDIES, and OCCIDENS) labeled at edge of dial plate instead of in compass bowl. Pin gnomon dial for Italian hours (labeled "die stund von Nidergang" [hours from sunset] and 10–23 in black numerals) and Babylonian hours (labeled "die stund von Auffgang" [hours from sunrise] and 1–14 in red numerals) .

IIb Undecorated. The maker's mark, a bird on a twig, appears once. Modern paper label reads "Troschel. 1600. Campbell 850.6." Image inverted for legibility.

12 Rectangular Ivory Diptych

signed "HANS TROSCHEL" [the Elder] and
maker's mark
dated 1611
Nuremberg, Germany
8.2 (w) x 12.2 (l) x 1.6 (h) cm
Formerly David P. Wheatland Collection
(purchased from E. Weil, 1950)
Inventory No. 7534

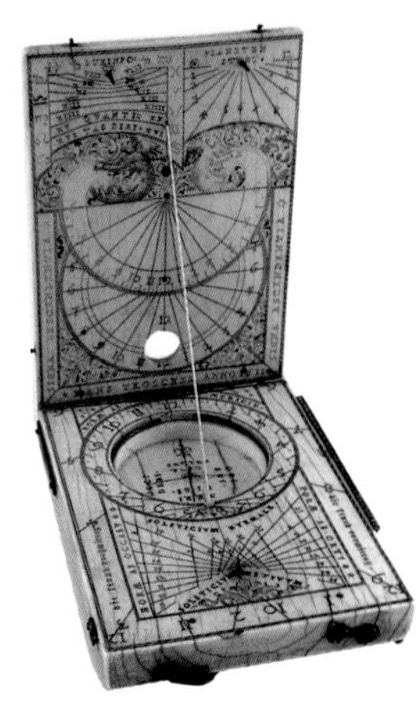

Faded black, red, and green coloring, with gilt accents. Compass viewing–hole in I; slot with brass clasp for gnomon rod (missing), brass arm to adjust relative angle of the two plates of ivory, brass scale to set latitude on sides of II; adjustment arm on I missing. Brass horse-head clasps on I and II, four bun feet on II. Compass needle and glass missing.

Ia Wind rose with 32 numbered points (east = 32), 16 labeled or abbreviated in German, hand-shaped brass index. Semicircular equinoctial dial labeled V–XII–VII in faded black/brown numerals for spring and summer months. Decorated with fruit and foliage.

Ib Pin gnomon dial labeled "QVANTITAS DIEI" [length of the day] and VIII–XVI twice in black numerals, marked with zodiacal symbols. Pin gnomon dial labeled "PLANETEN STUNDEN" [planetary hours] and 0–12 in black numerals. Semicircular vertical dial labeled 6–12–6 in red numerals, with lines and dots marking the half- and quarter-hours, respectively. Equinoctial dial labeled 5–12 –7 in black numerals for autumn and winter months. Engravings of two whimsical angels. Inscribed "HANS TROSCHEL ANNO 1611" [Hans Troschel, year 1611], "SIGNA DESCENDENTIA" and "SIGNA ASCENDENTIA" [descending and ascending signs].

IIa Horizontal dial for approximate latitude 55.5°, labeled 4–12–8 in red numerals, with lines and dots marking the half- and quarter-hours, respectively. Inscribed "HORAE AB MERIDIE" [hours from midday] in red. Additional hour lines (labeled "die stund vormittag" [hours before noon] and 6–12 in red, and "die stund nachmittag" [hours after noon] and

12–6 in black) around the edge of the dial: the shadow of the dial's edge determines the time when the upper plate is inclined to the proper angle. Compass bowl labeled with cardinal directions: SEPTENTRIO, ORIENS, MERIDIES, and OCCIDENS. Magnetic declination 5° east of north. Pin gnomon dial for Italian hours (labeled "HORAE AB OCCASV" [hours from sunset] and 9–23 in faded brown) and Babylonian hours (labeled "HORAE AB ORTV" [hours from sunrise] and 1–15 in red). Inscribed "SOLSTICIVM HYEMALE" and "SOLSTICIVM ESTIVVM" [winter and summer solstices] in black.

IIb Standard German lunar volvelle with an engraved and punched brass rotating disc (3.1 cm). Scales labeled 1–12 twice (inner and middle) and 1–29 (outer). Epact tables (labeled "EPACTA IVLIANII" and "EPACTA GREGORII" [Julian and Gregorian epacts] beginning with 4 and 24, respectively, for the year 1609) and correction table for lunar-solar conversion (in hours and minutes) tabulated in circular form around volvelle. Latitude table for 24 cities, most in Europe. Fruit decoration. Maker's mark (bird on a twig) at top of epact table. Image inverted for legibility.

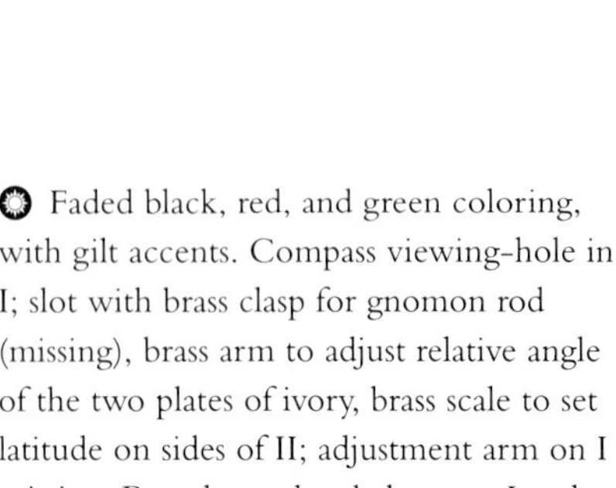

This elaborate diptych possesses several features worthy of note. The first is the inclusion of an unusual dial projection along the edges of face IIa. This auxiliary dial works as follows: the upper tablet is inclined to the proper angle (determined by the local latitude), with the metal arm and scale attached to the right side of the dial. When this tablet is level and oriented north-south (the hinge facing south), the shadow of the upper plate's edge will indicate the time along the hour scales. Note

that there are two hour scales — one from 6 a.m. to noon along the right side, the other from noon to 6 p.m. along the left — with the two scales overlapping along the shorter edge from 9 a.m. to 3 p.m.

Perhaps the first use of this projection is found on a dial by the Nuremberg mathematician Georg Hartmann (1489–1564) in the Max Elskamp collection (see Michel, 1953, no. 173), dated 1537. Other examples by Hans Troschel the Elder and Hans Troschel the Younger are known.

Also of interest on this diptych is the correction scale surrounding the lunar volvelle. This scale has two components, one for hours and one for minutes. The correction for the first day following the new moon (known as the "lateness" of the moon, or the difference between "local solar" and "local moon" times; see Section 1 for a further explanation) is 48 minutes, for the second day 1 hour 36 minutes, for the third day 2 hours 24 minutes, and so forth; the correction increases by 48 minutes daily until the full moon, after which it decreases by 48 minutes each day.

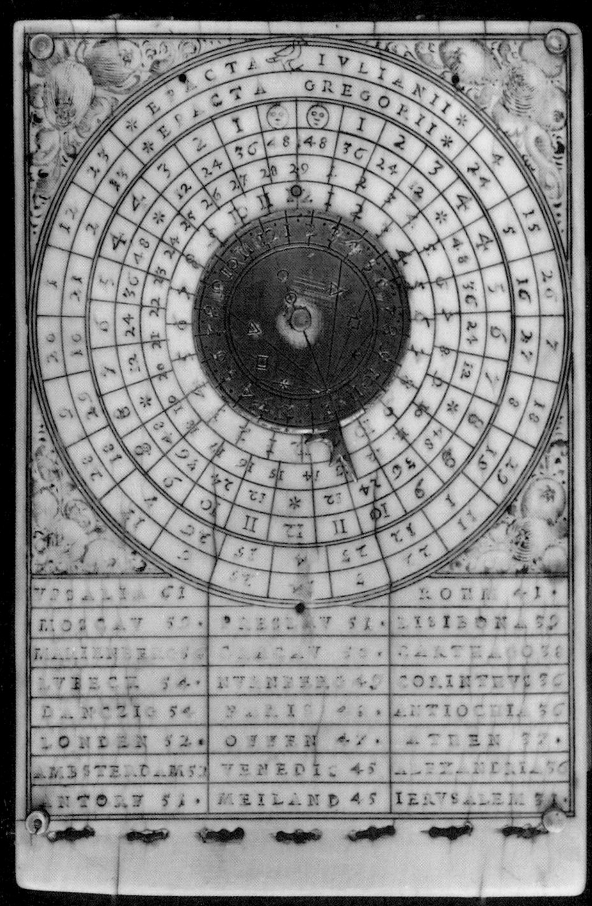
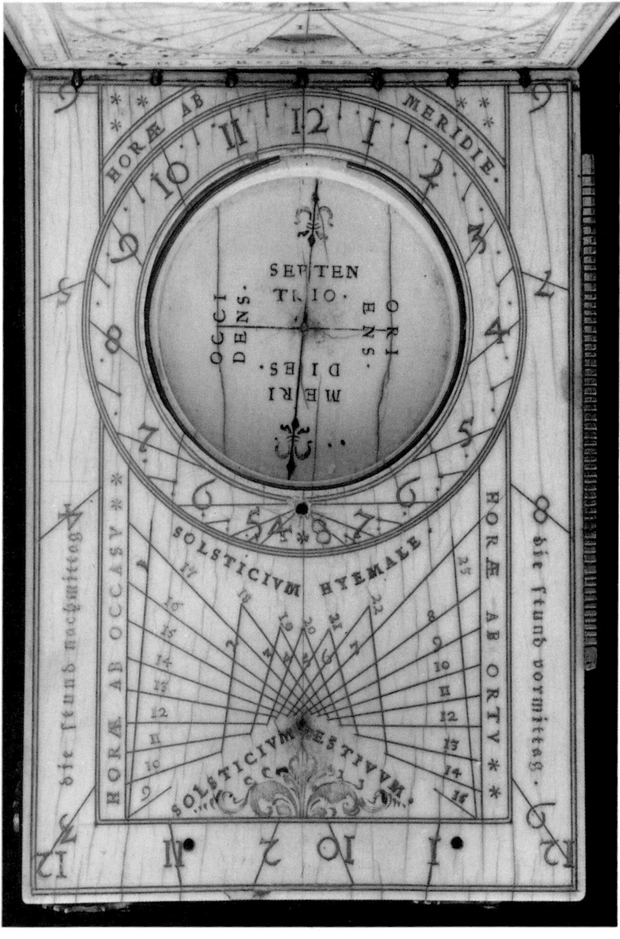

13 Rectangular Ivory Diptych

signed "HANS TROSCHEL" [the Younger]
and maker's mark
dated 1626
Nuremberg, Germany
6.3 (w) x 9.9 (l) x 1.4 (h) cm
Formerly Drecker Collection No. 123,
then David P. Wheatland Collection
Inventory No. 7458

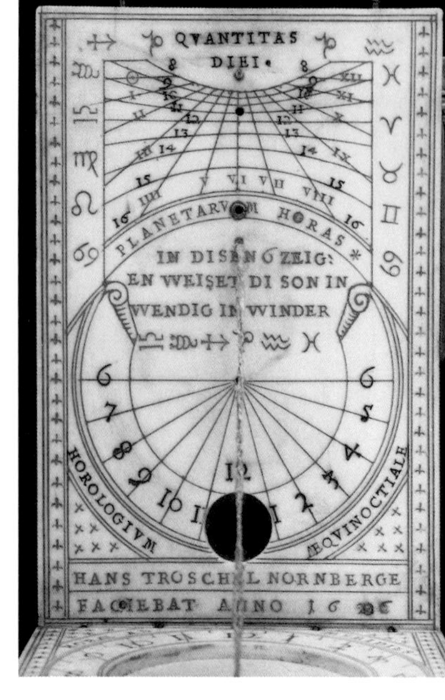

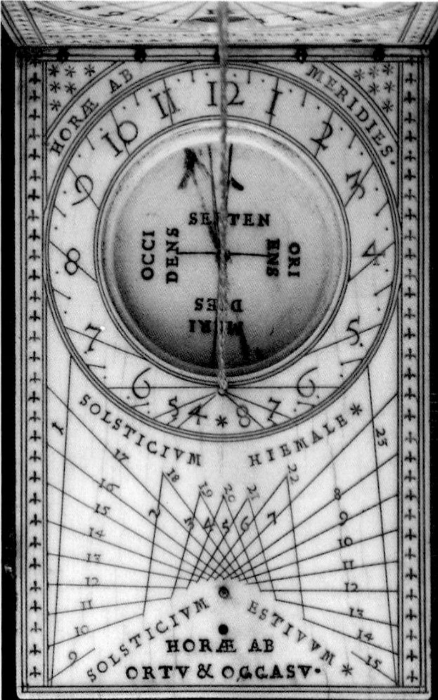

◉ Black, red, and green coloring. Compass viewing-hole with intact glass in I; slot with brass clasp for needle gnomon (missing), brass arm and scale to adjust for latitude on sides of II. Brass horse-head clasps on I and II, four bun feet on II. Compass needle and glass missing. Numbered 458 in red along outer edge near hinge.

Ia Wind rose with 32 numbered directions (east = 32), 16 labeled or abbreviated in German, hand-shaped index. Semicircular equinoctial dial for spring and summer months labeled VI–XII–VI in black numerals, with lines and dots marking the half- and quarter-hours, respectively, marked with red zodiacal symbols of spring and summer months. Inscribed "MONSTRO VIAM PERGE SECVRVS" [I show the way, proceed securely] at corners.

Ib Pin gnomon dial labeled "QVANTITAS DIEI" [length of the day] and 8–16 twice in black numerals, with lines marked "PLANETARVM HORAS" [planetary hours] labeled 0–XII in red numerals, marked with red zodiacal symbols of autumn and winter months. Equinoctial dial labeled "HOROLOGIVM AEQVINOCTIALE" [equinoctial sundial] and 6–12–6 in black numerals. Inscribed "HANS TROSCHEL NORNBERGE FACIEBAT ANNO 1626" [made by Hans Troschel of Nuremberg in the year 1626] and "IN DIESEN 6 ZEIGEN WEISET DI SON IN WENDIG IN WINDER" [in these six astrological signs the sun will indicate the time conveniently in winter].

IIa Horizontal dial with a single scale for approximate latitude 49°, labeled "HORAE AB MERIDIES" [hours from midday] and 4–12–8 in red numerals, with lines and dots marking the half- and quarter-hours, respectively. Cardinal directions indicated

in the compass bowl. Magnetic declination indicated for 5° east and 7° and 20° west of north. Pin gnomon dial for Italian hours (labeled 9–23 in gray numerals) and Babylonian hours (labeled 1–15 in red numerals) labeled "HORAE AB ORTV & OCCASV" [hours from sunrise and sunset]. Shadow lengths for the solstices labeled "SOLSTICIVM HIEMALE" and "SOSTICIVM ESTIVVM" [winter and summer solstices].

IIb Latitude table for 38 cities, most in Europe. Maker's mark (six-pointed star) appears twice near hinge. Image inverted for legibility.

REVELN	61	PARIS	47
VPSALIA	61	INSPRVCK	46
GOTTLANDIA	60	MEILAND	45
STOCKOLM	60	LEON	45
COPPENHAGEN	56	VENEDIG	44
DANCZIG	54	COMPOSTEL	44
HAMBVRG	54	MARSILIA	43
EMBDEN	53	FLORENCZ	43
AMSTERDAM	52	ROHM	41
BRANDENBVRG	52	TROIA	41
ANTORF	51	NEAPLES	40
BRESLAV	51	LYSIBONA	39
CRACAV	50	CARTHAGO	38
PRAG	50	ATHEN	37
HEIDELBERG	49	EPHESVS	37
STRASBVRG	48	ALEXANDRIA	36
NVRNBERG	49	CORINTHVS	36
WIEN	48	NINEVE	36
OFEN	47	HIERVSALEM	31

14 Rectangular Ivory Diptych

signed "HANNS TROSCHEL" [the Younger]
and maker's mark
before 1634
Nuremberg, Germany
5.4 (w) x 8.9 (l) x 1.4 (h) cm
Formerly Drecker Collection No. 124,
then David P. Wheatland Collection
Inventory No. 7535

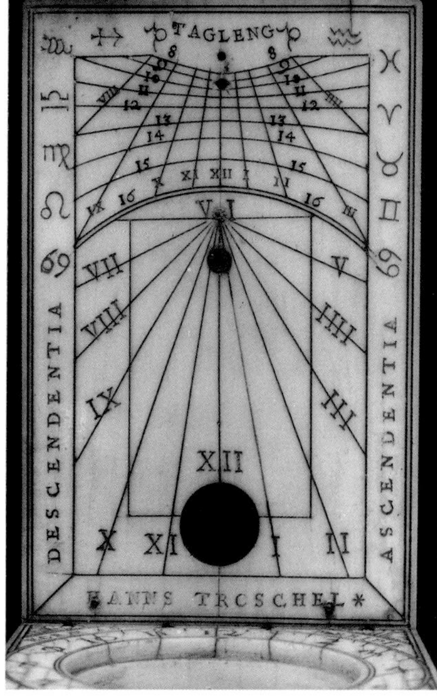

Ia | Ib
IIb | IIa

⊙ Faded black and red coloring. Compass viewing-hole in I. Brass wire clasps on I and II, four bun feet on II. Compass needle and glass missing.

Ia Wind rose with 32 numbered directions (east = 32), 16 labeled or abbreviated in German, hand-shaped index. Inscribed "MONSTRO VIAM PERGE SECVRVS" [I show the way, proceed securely]. Single-lined border.

Ib Pin gnomon dial labeled "TAGLENG" [length of the day], with hour lines for common hours (labeled VIII–XII–IIII in red numerals) and the length of the day (labeled 8–16 twice in black numerals), marked with red zodiacal symbols. On the left the symbols are labeled "DESCENDENTIA" and on the right "ASCENDENTIA" [descending and ascending astrological signs]. Vertical dial labeled VI–XII–VI in red numerals. Signed "HANNS TROSCHEL." Double-lined border.

IIa Horizontal dial with a single scale for approximate latitude 47.5°, labeled "HORAE AB MERIDIE" [hours from midday] and 4–12–8 in red numerals, with lines and dots marking the half- and quarter-hours, respectively. Cardinal directions indicated in the compass bowl: SEPTEN, ORIENS, MERIDIES, and OCCIDENS. Magnetic declination 5° east of north. Pin gnomon dial for Italian hours (labeled 9–23 in gray numerals) and Babylonian hours (labeled 1–15 in red numerals) labeled "HORAE AB ORTV ET OCCASV" [hours from sunrise and sunset]. Double-lined border.

IIb Standard German lunar volvelle with a stamped rotating brass disc (2.1 cm) with index and man-in-the-moon motif. Epact scales (labeled "EPACTA IVLIANII" and "EPACTA GREGORI" [Julian and Gregorian

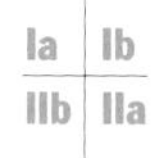

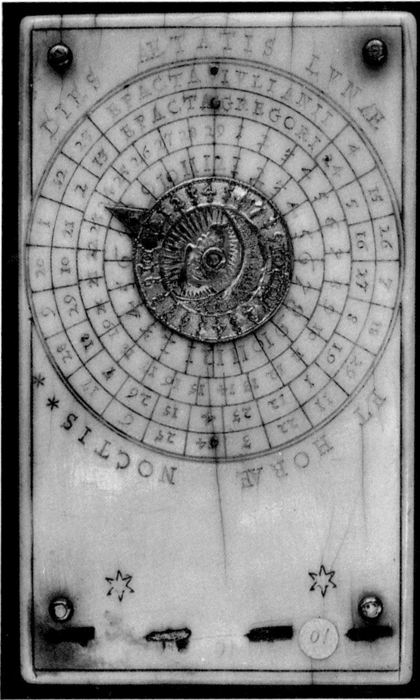

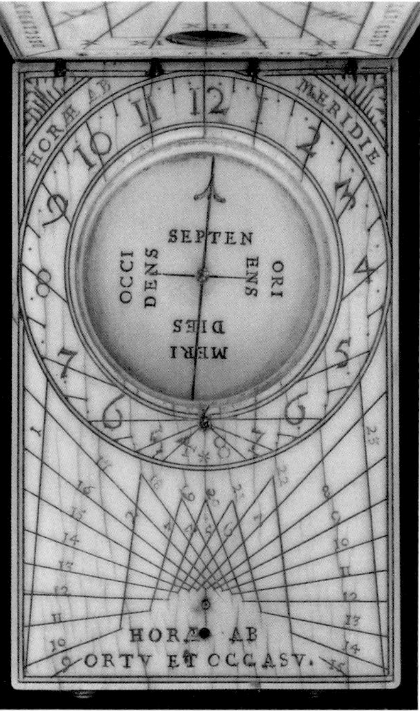

epacts] and beginning with 4 and 24, respectively, for the year 1628) tabulated in circular form around volvelle. Inscribed "DIES AETATIS LVNAE ET HORAE NOCTIS" [days of the age of the moon and night hours]. Maker's mark (six-pointed star) appears twice near the hinge. Small round paper label with the number 10 in red, and again in red (very faint) along the hinged edge. Single-lined border. Image inverted for legibility.

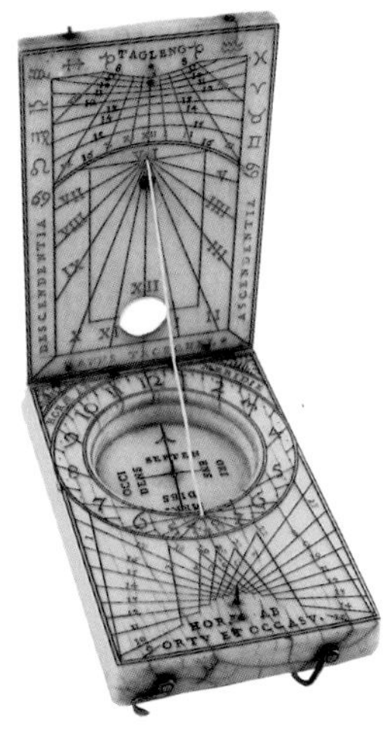

15 Oval Ivory Diptych

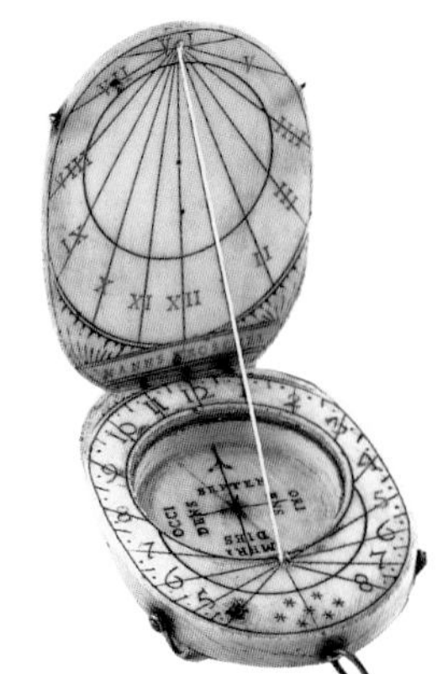

signed "HANNS TROSCHEL" [the Younger]
before 1634
Nuremberg, Germany
5.1 (w) x 7.2 (l) x 1.2 (h) cm
Formerly Drecker Collection No. 125,
then David P. Wheatland Collection
Inventory No. 7532

⬢ Black and red coloring. Brass wire clasps on I and II, hanging loop on II. Both plates are slightly warped. The wire hinge has become detached from I. While the outer covers of this dial are highly decorated, the sundials themselves (Ib and IIa) are very plain. Compass needle and glass missing.

Ia Engraving in black of two figures amidst the clouds passing a lighted torch. Also present is an angel floating above a shining sun.

Ib Vertical dial labeled VI–XII–VI in red numerals. "HANNS TROSCHEL" is inscribed beneath the dial.

IIa Horizontal dial with a single scale for approximate latitude 50.5°, labeled 4–12–8 in red numerals, with lines and dots marking the half- and quarter-hours, respectively. Cardinal directions labeled SEPTEN, ORIENS, MERIDIES, and OCCIDENS. Magnetic declination indicated at 5° east of north, and also faintly at 21° west. Decorated only with seven red asterisk stamps.

IIb Engraving in black of two symbolic figures on a dais. The left figure, Justice, holds a set of scales and wears a crown and an eye on her breast. The right figure, Truth, holds a palm leaf and a smiling sun and places her foot upon an orb (see Ripa, 1971, no. 50, 120). Image inverted for legibility.

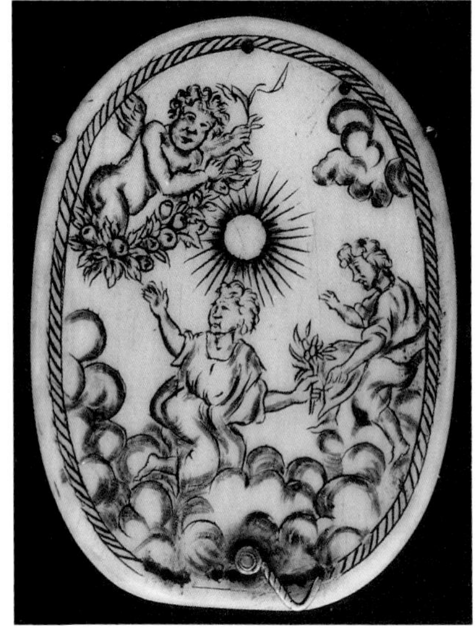

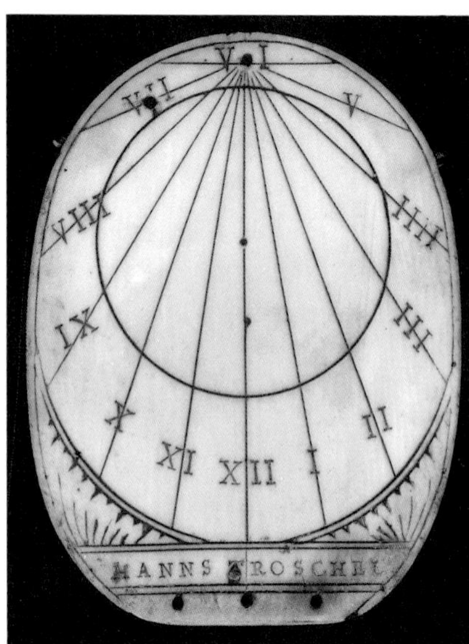

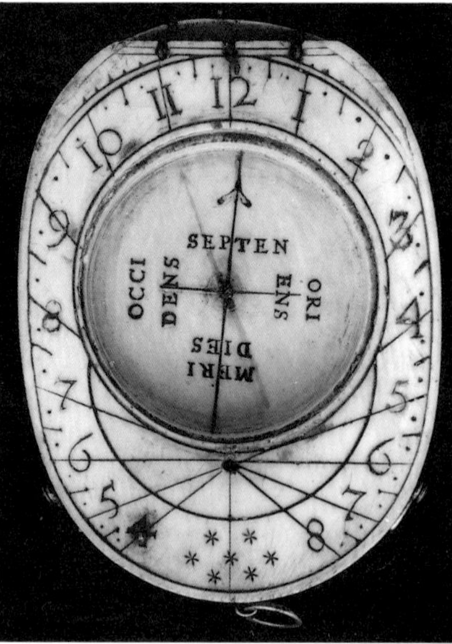

<table>
<tr><td>Ia</td><td>Ib</td></tr>
<tr><td>IIb</td><td>IIa</td></tr>
</table>

The Miller Family Workshop

The output of the Miller workshop, more than any other family's products, reveals how it is probably misleading to ascribe an instrument to an
individual maker rather than to a workshop. All of the Miller dials in this
collection are signed (or at least initialed) Lienhart Miller, Leonhart Miller,
or Nicolaus Miller. One finds diptychs in the identical style and format
signed by Lienhart and by Leonhart Miller. Were these the signatures of
two separate makers, or were they simply variant spellings of the same
name? On the one hand, one finds spellings of the family name as "MILEL"
(cat. no. 17), "MILER" (cat. no. 22), "MIELER," "MILRE," and "MNLLER."
Also, the archival information appears to provide evidence for a single
Kompassmacher who died in 1653, although further research needs to be
conducted; lack of archival information does not necessarily preclude the
existence of two persons bearing the same name. On the other hand, a
simple analysis of the available diptychs that are both signed and dated
yields the following: dials made between 1613 and 1629 are consistently
signed "LIENHART" (with only one exception, a 1625 dial with the
spelling "LINHART"), while dials produced between 1630 and 1649 are
always signed "LEONHART" (see Appendix II). A continuous output from
1613 to 1649 is not inconceivable, but possibly we have here two different
Kompassmachers. A series of four nearly identical fleur-de-lys maker's
marks are found on dials signed Lienhart and Leonhart. Furthermore, although Leonhart died in 1653, this mark continued to be used on unsigned
Miller-style diptychs, presumably produced in the Miller workshop,
through the 1670s.

Dated diptychs attributed to Nicolaus Miller indicate that he was
active in the years 1647–1661. He adopted a crown as his maker's mark,
similar to but distinguishable from the ones used by Paul Reinmann and
Michael Lesel, rather than adopting his family's fleur-de-lys.

The Miller workshop produced two "standard models" shown in this
catalogue. The fancier of the two, referred to in the descriptions as Miller
type 1 (cat. nos. 16, 18, and 20), examples of which were signed both
Lienhart and Leonhart, is typified by a slightly larger overall size than
other diptychs produced in the workshop (both larger and thicker tablets
of ivory), bright colors (including green and a rusty brown/orange), one
or more scaphe pin gnomon dials, and elaborate foliate decoration. These
stylistic elements are also found on the dials produced earlier by Michael
Lesel and later by Joseph Ducher. This "model" includes extensive latitude
lists, which were functional for latitudes between 39 degrees (in southern
Italy and Spain) and 54 degrees (in northern Germany).

A smaller "standard model," referred to in the descriptions as Miller
type 2 (cat. nos. 17, 19, 21–23), with a few minor variations, was produced in large quantities for over thirty years in the Miller workshop.
Dials of this sort were signed Lienhart, Leonhart, and Nicolaus Miller. A
common motto on these dials is the Latin phrase "SOLI DEO GLORIA,"
which in the context of a sundial could have a dual meaning. It can be
translated as "glory to God alone" (which would be the standard translation), or "glory to God, the sun," which would be an especially appro-

Miller family
maker's marks:

Lienhart or
Leonhart

Lienhart or
Leonhart

Nicolaus

priate aphorism to place on a sundial. A common motif on this "model" of diptych is that of small meteors or comets at the bottom of face Ib. Since these dials were continuously manufactured over such a long period, there appears to be no correspondence between the dates on the dials so ornamented with documented reports of comets or anomalous meteor showers.

Both models share common stylistic elements, most notably the use of foliate decoration. The use of lines, dots, and stars to indicate the half- and quarter-hours is also standard on Miller diptychs.

The Miller workshop, like the Karners', also produced composite diptychs, composed of thin slices of ivory on either side of a wooden core. Beginning in the mid-seventeenth century the production of composite dials increased, perhaps in response to changes in the availability and cost of larger pieces of ivory. Nicolaus Miller introduced another "standard model," the octagonal composite diptych (cat. nos. 24 and 25). Such dials were presumably cheaper to produce.

16 Rectangular Ivory Diptych

signed "LIENHART MILLER" and
maker's mark
dated 1613
Nuremberg, Germany
7.0 (w) x 10.4 (l) x 1.2 (h) cm
Formerly Drecker Collection No. 126,
then David P. Wheatland Collection
Inventory No. 7459

◉ Black, red, brown, and green coloring. Miller type 1 diptych. Compass viewing-hole in I, 2 slots with brass clasps for wind vane and quadrant pendant in side of II. Brass horse-head clasps and 4 bun feet on each I and II. Compass needle intact, but no glass.

Ia Wind rose with 16 points labeled or abbreviated in German, 32 numbered directions (east = 32), and 8 weather indicators labeled in German: WARM HEITER, SCHON MITELMESIG, WARM FEICHT, RENGISCH, KALT FEICHT, SCHNEIG, SCHON DRVCKEN, and HEITER KALT. Hand-shaped indicator with crescent moon-face at the other end. Wind vane missing. Sun-face motif at center of wind rose. Foliate decoration.

Ib Pin gnomon dial (gnomon missing) labeled "QVANTITAS DIEI" [length of the day], labeled 8–16 in red numerals, and marked with zodiacal symbols. Holes for string gnomon at 39°, 42°, 45°, 48°, 51°, and 54° latitude. Latitude table for 30 cities (39°–54° latitude). Foliate decoration.

IIa Horizontal dial with six scales for latitudes 39° (inner), 42°, 45°, 48°, 51°, and 54° (outer), labeled 4–12–8 in black numerals in outermost scale. Scaphe dial with sun-face motif for Italian hours (labeled 8–23 in black numerals) and Babylonian hours (labeled 1–16 in red numerals). Scaphe dial is inscribed "LIENHART MILLER 1613." Compass bowl with faint fleur-de-lys maker's mark to left of compass pivot. Cardinal directions labeled SEPT, ORIE, MERI, and OCCI. Magnetic declination 2° east of north. Foliate decoration with two long-tongued squirrels.

IIb Quadrant for measuring elevations, numbered 5°–90° by fives, with 1° increments. Colorful engraving of a forest scene

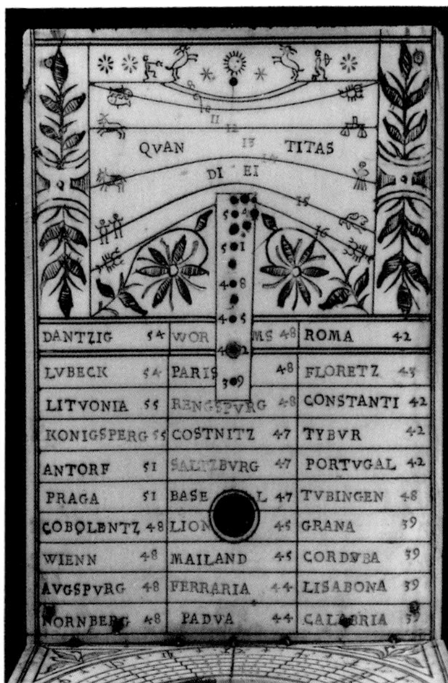

DANTZIG	54	WOR MS	48	ROMA	42
LVBECK	54	PARIS	48	FLORETZ	43
LITVONIA	55	RENGSPVRG	48	CONSTANTI	42
KONIGSPERG	55	COSTNITZ	47	TYBVR	42
ANTORF	51	SALTZBVRG	47	PORTVGAL	42
PRAGA	51	BASEL	47	TVBINGEN	48
COBOLENTZ	48	LION	45	GRANA	39
WIENN	48	MAILAND	45	CORDVBA	39
AVGSPVRG	48	FERRARIA	44	LISABONA	39
NORNBERG	48	PADVA	44	CALABRIA	39

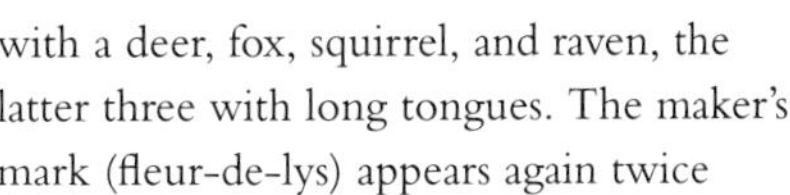

with a deer, fox, squirrel, and raven, the latter three with long tongues. The maker's mark (fleur-de-lys) appears again twice (faintly) near the hinge. Image inverted for legibility.

17 Rectangular Ivory Diptych

signed "LIENHART MILEL" and
maker's mark
dated 1616
Nuremberg, Germany
5.6 (w) x 9.1 (l) x 1.1 (h) cm
Formerly David P. Wheatland Collection
(purchased in France, 1952)
Inventory No. 7565

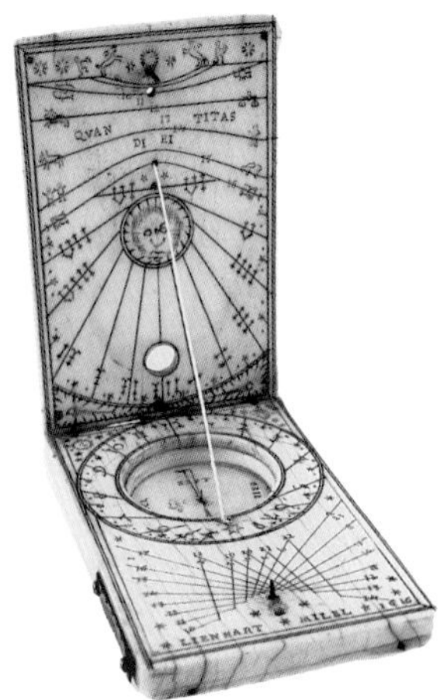

⚙ Black and red coloring. Miller type 2 diptych. Leaf I is chipped slightly at the hinge, and the brass hinge clasp is broken in half. Compass viewing-hole in I, slot with brass clasp for wind vane in side of II. Four brass bun feet on both I and II. Compass needle, glass, and metal ring missing.

Ia Wind rose with 16 directions labeled in German, wind vane missing. Sun-face motif at center of wind rose. Foliate decoration.

Ib Pin gnomon dial labeled "QVANTITAS DIEI" [length of the day] and 8–16 in red numerals, marked with zodiacal symbols. Vertical dial labeled VI–XII–VI in black Gothic-style Roman numerals. Sun-face motif. Small decorative stamps and two engravings of a comet or a meteor.

IIa Horizontal dial with a single scale for approximate latitude 51°, labeled 4–12–8 in black numerals. Pin gnomon dial for Babylonian hours (labeled 1–14 in red numerals) and Italian hours (labeled 10–23 in black numerals). Compass bowl with fleur-de-lys maker's mark. Cardinal directions in compass bowl labeled SEPT, ORIE, MERI, and OCCI. Magnetic declination indicated at 7° east and 1° west of north. Sun and moon decorations. Small decorative stamps. Signed "LIENHART MILEL 1616."

IIb Standard German lunar volvelle, brass disc missing. Scales labeled 1–12 twice (inner) and 1–29 (outer). Foliate decoration. Maker's mark appears once near hinge (very faint).

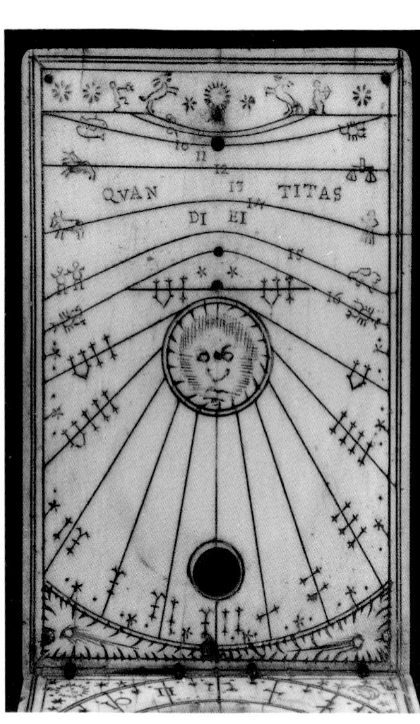

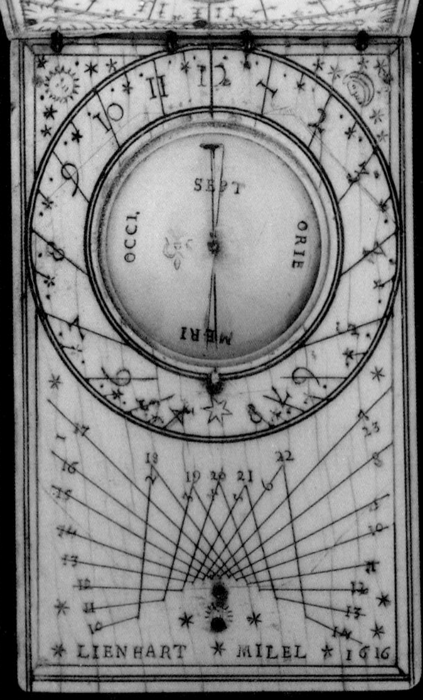

Ia	Ib
IIb	IIa

18 Rectangular Ivory Diptych

signed "LIENHART MILLER" and
maker's mark
dated 1629
Nuremberg, Germany
6.9 (w) x 11.1 (l) x 1.3 (h) cm
Formerly Drecker Collection No. 128,
then David P. Wheatland Collection
Inventory No. 7560

⬤ Black, red, brown, and green coloring.
Miller type 1 diptych. Compass viewing-
hole in I, slot with brass clasp for wind
vane in side of II. Brass horse-head clasps
on I and II and 4 bun feet on II. Compass
needle and glass missing. Paper label along
left outer edge reads "dr. 128."

Ia Wind rose with 16 points labeled or
abbreviated in German, 32 numbered
directions (east = 32) and 8 wind directions
labeled in Italian: PASSATVMO, SIROCHO,
OSTRO, LONENTE, PONENTE, MASTRO,
TRAMANTAM, and GREGORIO. Arrow-
shaped indicator. Wind vane missing. Sun
motif at center of wind rose. Four-winds
and stellar decoration.

Ib Pin gnomon dial labeled "QVANTITAS
DIEI" [length of the day], labeled 8–16 in
red numerals and marked with zodiacal
symbols. Holes for string gnomon at 42°,
45°, 48°, 51°, and 54° latitude. Engraving
of a village. Latitude table of 26 cities
(42°– 54° latitude). Foliate decoration.

IIa Horizontal dial with five scales for
latitudes 42° (inner), 45°, 48°, 51°, and 54°
(outer), labeled 4–12–8 in black numerals
in outermost scale. Inscribed "SOLI DEO
GLORIA" [glory to God alone]. Cardinal
directions in compass bowl labeled SEPT,
ORIE, MERI, and OCCI. Four wind-faces in
compass bowl. Magnetic declination 4°
east of north. Scaphe dial with sun-face
motif for Italian hours (labeled 8–23 in
black numerals) and Babylonian hours
(labeled 1–16 in red numerals). Inscribed
"LIENHART MILLER 1629" in scaphe dial.
Foliate decoration with various fruits.

IIb Standard German lunar volvelle with
an engraved and punched rotating brass
disc (3.4 cm) with index, marked with
astrological aspects. Scales labeled 1–12
twice (inner and middle) and 1–29 (outer).
Outer circle inscribed "DER NEA MAN,"
"DAS ERST VIERTEL," "DER VOL MAN," and
"DAS LECZT VIERTEL" [the new moon, the
first quarter, the full moon, and the last
quarter] with depictions of lunar phases.
Initials "LM" with maker's mark (fleur-de-
lys) between initials appear twice near
hinge. Colorful foliate decoration with
various fruits. Image inverted for legibility.

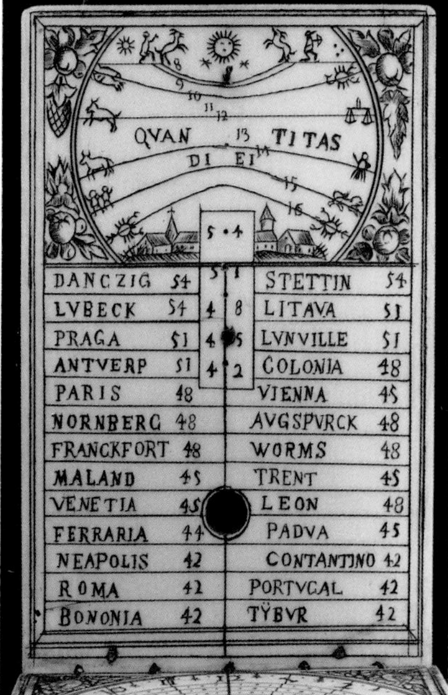

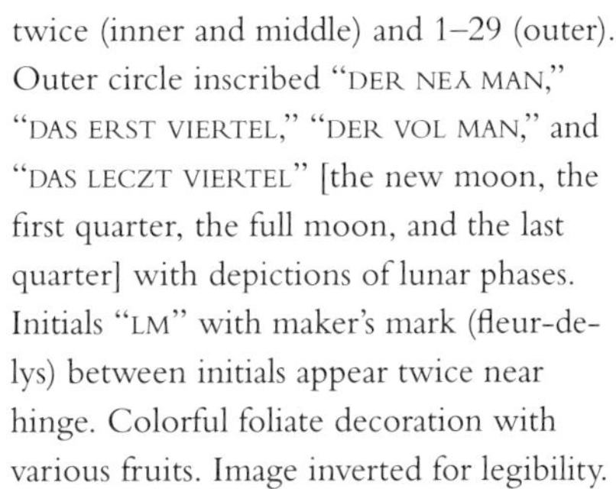

19 Rectangular Ivory Diptych

signed "LEONHART MILLER" and
maker's mark
dated 1630
Nuremberg, Germany
5.6 (w) x 9.1 (l) x 1.1 (h) cm
Formerly Harold Gillingham Collection
No. 84, then David P. Wheatland
Collection
Inventory No. 7567

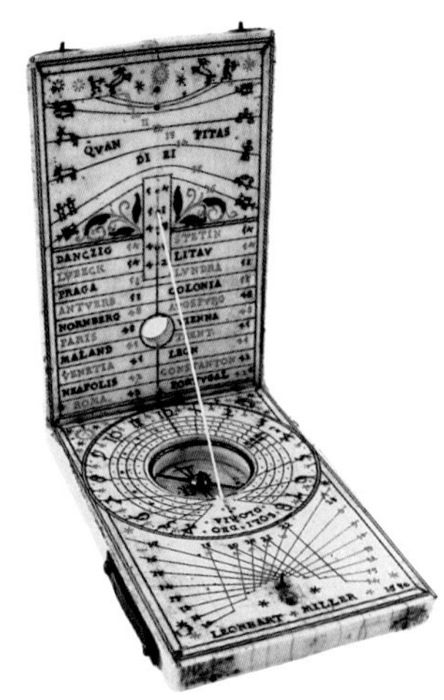

⬤ Red and black coloring. Miller type 2 diptych. Compass viewing-hole in I, slot with brass clasp for wind vane in side of II. Brass horse-head clasp on I, four bun feet on each I and II. Compass needle and brass ring intact, but glass missing. Paper label along left outer edge reads "AUSTRIA. INNSBRUCK. 1630. (M.R.W. 1930.)."

Ia Wind rose with 8 points, 16 directions labeled in German, wind vane missing. Sun-face motif at center of wind rose. Foliate decoration.

Ib Pin gnomon dial labeled "QVANTITAS DIEI" [length of the day] and 8–16 in bright red numerals, marked with zodiacal symbols. Holes for string gnomon at 42°, 45°, 48°, 51°, and 54°. Latitude table for 20 European cities (42°–54° latitude). Foliate decoration. Small decorative stamps.

IIa Horizontal dial with five scales for latitudes 42° (inner), 45°, 48°, 51°, and 54° (outer), labeled 4–12–8 in black numerals in outermost scale. Inscribed "SOLI DEO GLORIA" [glory to God alone]. Compass with fleur-de-lys maker's mark. Cardinal directions given in compass bowl as SEPT, ORIE, MERI, and OCCI. Magnetic declination indicated at 7° east and 1° and 10° west of north. Pin gnomon dial for Babylonian hours (labeled 1–14 in bright red numerals) and Italian hours (labeled 10–23 in black numerals). Sun and moon decorations and several small stamps. Signed "LEONHART MILLER 1630."

IIb Standard German lunar volvelle, brass disc missing. Scales labeled 1–12 twice (inner) and 1–29 (outer). Initials "LM" with maker's mark (fleur-de-lys) between initials appear twice near hinge. Foliate decoration. Image inverted for legibility.

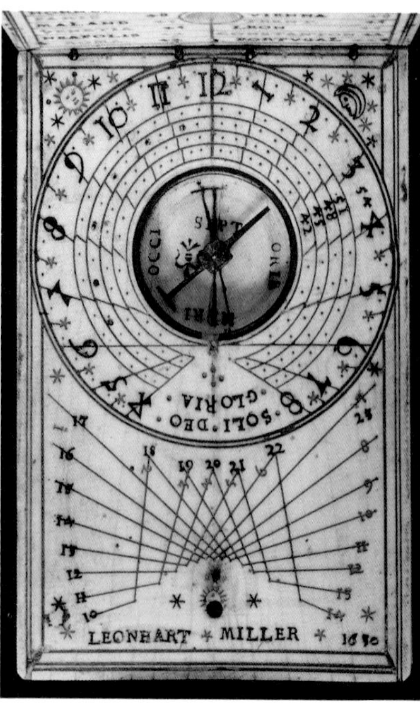

20 Rectangular Ivory Diptych

signed "LEONHART MILLER" and
maker's mark
dated 1636
Nuremberg, Germany
7.0 (w) x 10.8 (l) x 1.4 (h) cm
Formerly David P. Wheatland Collection
(purchased from R. A. T. Lier, 1950)
Inventory No. 7568

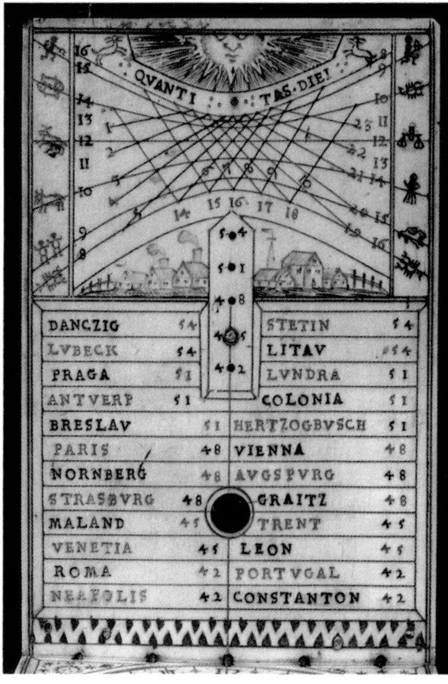

⊛ Black, red, and green coloring. Miller type 1 diptych. Compass viewing-hole in I, slot with brass clasp for wind vane in side of II. Brass horse-head clasps on I and II. Four bun feet on I and on II. Compass needle, glass, and metal ring intact.

Ia Wind rose with 16 points labeled or abbreviated in German, 32 numbered directions (east = 32), and 8 directions labeled in Italian. Brass arrow indicator, wind vane missing. Sun-face motif at center of wind rose. Four-winds decoration in corners. Elaborate colorful engraving.

Ib Pin gnomon dial labeled "QVANTITAS DIEI" [length of the day], labeled 16–8 in black numerals and 8–16 in red numerals, marked with zodiacal symbols and hour lines for Italian hours (labeled 14–23 in red numerals) and Babylonian hours (labeled 1–10 in green numerals). Engraving of the sun and a village. Holes for string gnomon at 42°, 45°, 48°, 51°, and 54°. Latitude list of 24 European cities (42°–54° latitude).

IIa Horizontal dial with five scales for latitudes 42° (inner), 45°, 48°, 51°, and 54° (outer), labeled 4–12–8 in black numerals in outermost scale. Inscribed "GLORIA SOLI DEO" [glory to God alone]. Scaphe dial with sun-face motif for Italian hours (labeled 8–23 in black numerals) and Babylonian hours (labeled 1–16 in red numerals). Signed "LEONHART MILLER 1636" in scaphe dial. Compass bowl with four-winds motif. Cardinal directions labeled in compass bowl. Magnetic declination indicated at 5° east and 17° west of north. Sun and moon decorations. Small stamps. Engraving of fanciful animals with green wings and faces with green foliage.

IIb Standard German lunar volvelle with an engraved and punched rotating brass

disc (3.6 cm) with index, marked with astrological aspects. Scales labeled 1–12 twice (inner and middle) and 1–29 (outer). Outermost scale inscribed "DER NEY MON," "DAS ERST VIERTEL," "DER VOL MON," and "DAS LETZ VIERTEL" [the full moon, the first quarter, the new moon, and the last quarter] with depictions of lunar phases. Initials "LM" with maker's mark (fleur-de-lys) between initials appear once near hinge. Colorful engraving of unusual vegetation. Image inverted for legibility.

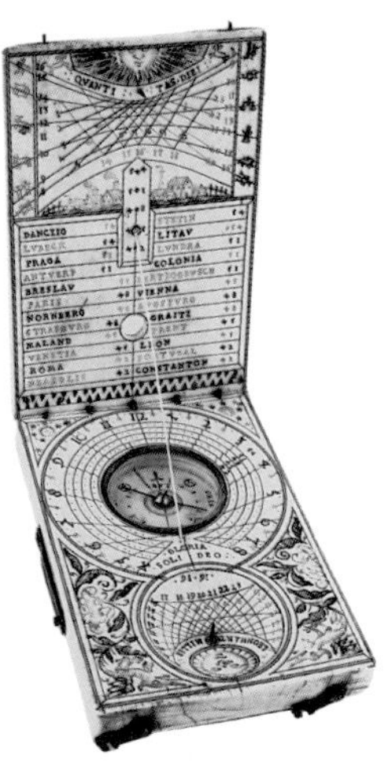

21 Rectangular Ivory Diptych

signed "LEONHART MILLER" and
maker's mark
dated 1640
Nuremberg, Germany
5.5 (w) x 9.1 (l) x 1.5 (h) cm
Formerly Drecker Collection No. 127,
then David P. Wheatland Collection
Inventory No. 7562

Faded black and red coloring. Miller type 2 diptych. Compass viewing-hole in I, slot for wind vane in side of II. All brass clasps and bun feet on I and II missing, as well as pin gnomon. Compass needle is a replacement. Glass and metal ring missing.

Ia Wind rose with 16 points labeled in German, index and wind vane missing. Sun-face motif at center of wind rose. Foliate decoration.

Ib Pin gnomon dial (gnomon missing) labeled "QVANTITAS DIEI" [length of the day] and 8–16 in faint red numerals, marked with zodiacal symbols. Holes for string gnomon at 42°, 45°, 48°, 51°, and 54° latitude. Latitude table for 20 European cities (42°–54° latitude). Foliate decoration.

IIa Horizontal dial with five scales for latitudes 42° (inner), 45°, 48°, 51°, and 54° (outer), labeled 4–12–8 in black numerals. Inscribed "SOLI DEO GLORIA" [glory to God alone]. Pin gnomon dial (gnomon missing) for Italian hours (labeled 10–23 in black numerals) and Babylonian hours (labeled 1–14 in red numerals). Compass bowl with fleur-de-lys maker's mark. Cardinal directions given in compass bowl as SEPT, ORIE, MERI, and OCCI. Magnetic declination 4° east of north. Sun and moon decorations and various other decorative stamps. Signed "LEONHART MILLER 1640."

IIb Standard German lunar volvelle, brass disc missing. Scales labeled 1–12 twice (inner) and 1–29 (outer). Initials "LM" with maker's mark (fleur-de-lys) in between appear twice near hinge. Foliate decoration. Numbered 32 on modern paper label. Image inverted for legibility.

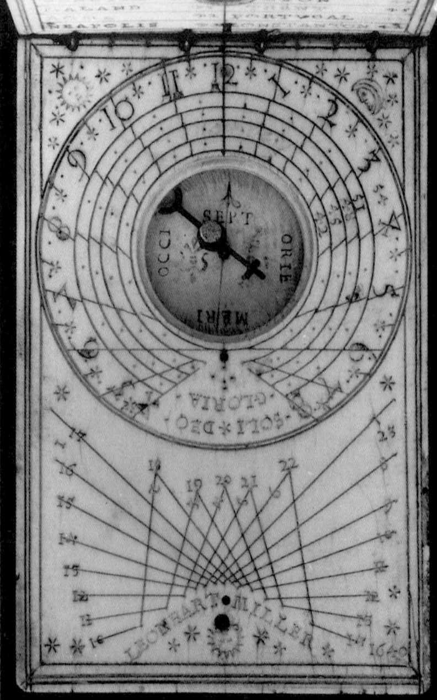

Ia | Ib
IIb | IIa

22 Rectangular Ivory Diptych

signed "LEONHART MILER" and
maker's mark
dated 1644
Nuremberg, Germany
5.1 (w) x 8.2 (l) x 1.2 (h) cm
Formerly Harold Gillingham Collection
No. 85, then David P. Wheatland
Collection
Inventory No. 7563

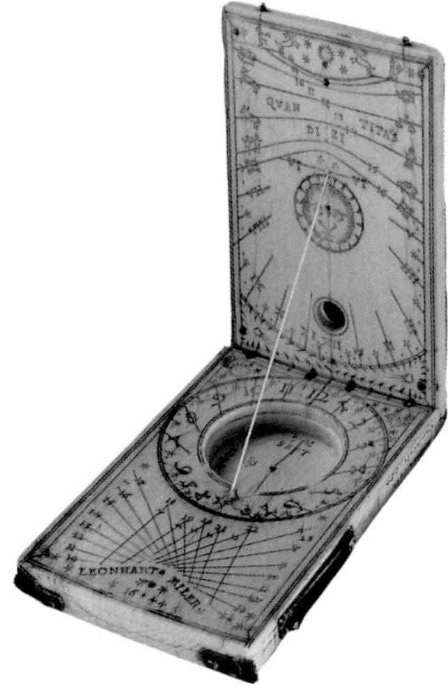

● Black and red coloring. Miller type 2 diptych. Leaf II slightly curved, beveled along longer edges. Compass viewing-hole in I, hole with brass clasp in II for wind vane. Brass horse-head clasps on I and II, four bun feet on I. Compass needle, glass, and metal ring missing.

Ia Wind rose with 8 points, 16 directions labeled in German. Wind vane missing. Sun-face motif at center of wind rose. Foliate decoration.

Ib Pin gnomon dial (gnomon missing) labeled "QVANTITAS DIEI" [length of the day] and 8–16 in rusty brown numerals, marked with zodiacal symbols. Vertical dial labeled VI–XII–VI in rusty brown numerals. Sun-face motif. Small decorative stamps. Engraving of two comets or meteors at bottom.

IIa Horizontal dial with a single scale for approximate latitude 48°, labeled 4–12–8 in black numerals. Compass bowl with fleur-de-lys maker's mark. Cardinal directions given in compass bowl as SEPT, ORIE, MERI, and OCCI. Magnetic declination not indicated: north-south line with arrowhead. Pin gnomon dial (gnomon missing) for Italian hours (labeled 10–23 in black numerals) and Babylonian hours (labeled 1–14 in red numerals). Sun and moon decorations. Small decorative stamps. Signed "LEONHART MILER 1644."

IIb Standard German lunar volvelle, brass disc missing. Scales labeled 1–12 twice (inner) and 1–29 (outer). Maker's mark (fleur-de-lys) appears twice near hinge (faint). Foliate decoration. Image inverted for legibility.

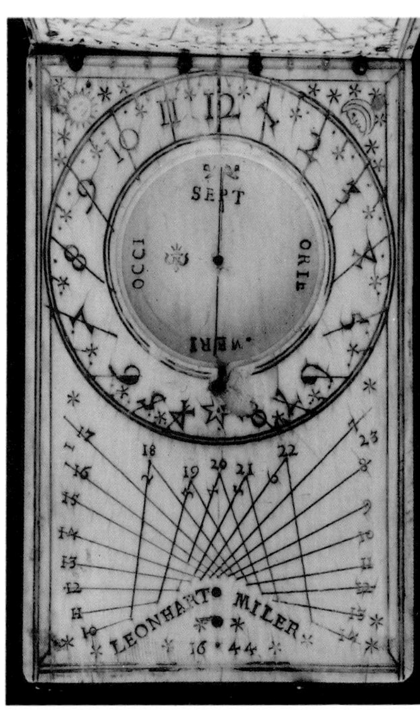

Ia | Ib
IIb | IIa

23 Rectangular Ivory Diptych

signed "NICOLAVS MILLER" and
maker's mark
dated 1649
Nuremberg, Germany
5.6 (w) x 9.1 (l) x 1.3 (h) cm
Formerly David P. Wheatland Collection
(purchased in France, 1952)
Inventory No. 7570

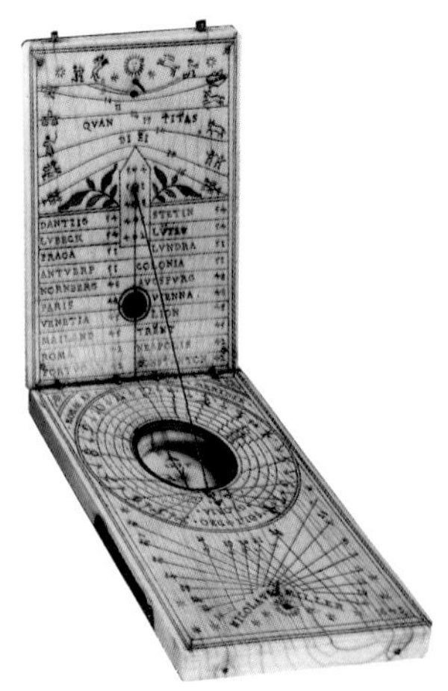

Black and red coloring. Miller type 2 diptych. Compass viewing-hole in I, slot with brass clasp for wind vane in side of II. Broken brass clasp on I, clasps on II missing, several bun feet on I (5) and II (4). Compass needle and glass missing. Brass ring intact.

Ia Wind rose with 16 points labeled in German, wind vane missing. Sun-face motif at center of wind rose. Foliate decoration.

Ib Pin gnomon dial labeled "QVANTITAS DIEI" [length of the day] and 8–16 in red numerals, marked with zodiacal symbols. Holes for string gnomon at 42°, 45°, 48°, 51°, and 54° latitude. Latitude table of 20 European cities (42°–54° latitude). Foliate decoration. Sun-face and other stamps.

IIa Horizontal dial with five scales for latitudes 42° (inner), 45°, 48°, 51°, and 54° (outer), labeled 4–12–8 in black numerals. Inscribed "SOLI DEO GLORIA" [glory to God alone] and "HORAE AB MERIDIE" [hours from midday]. Pin gnomon dial for Italian hours (labeled 10–23 in black numerals) and Babylonian hours (labeled 1–14 in red numerals). Compass bowl with crown maker's mark. Cardinal directions given in compass bowl as SEPT, ORI, MERI, and OCCI. Magnetic declination 3° east of north. Decorative stamps. Signed "NICOLAVS MILLER 1649."

IIb Standard German lunar volvelle, brass disc missing. Scales labeled 1–12 twice (inner) and 1–29 (outer). Initials "NM" with maker's mark (crown) in between appear once near hinge. Foliate decoration. Image inverted for legibility.

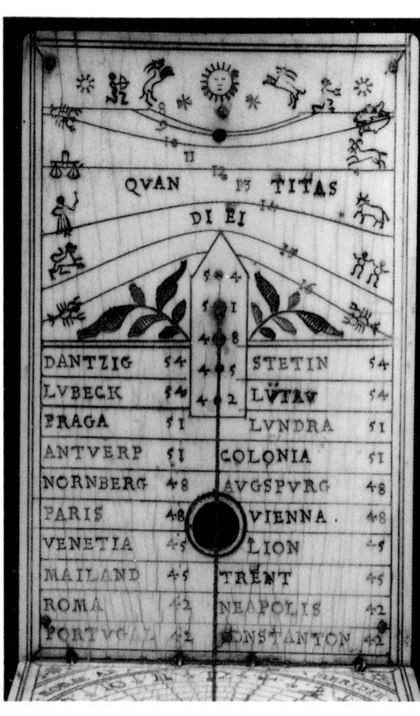

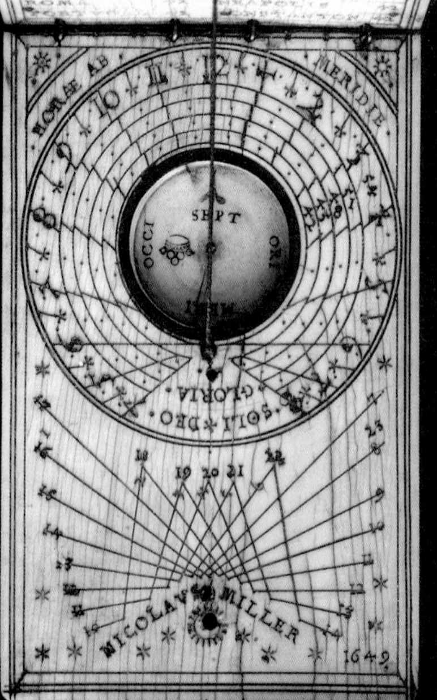

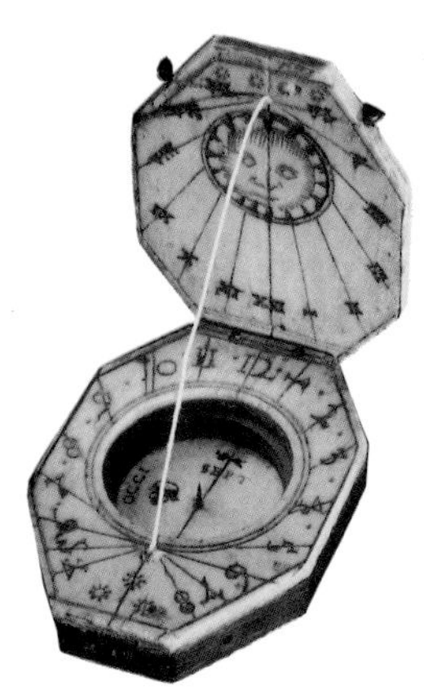

24 Octagonal Ivory and Wood Diptych

signed "NM" [Nicolaus Miller] and
maker's mark
c. 1650
Nuremberg, Germany
3.9 (w) x 5.4 (l) x 1.2 (h) cm
Formerly David P. Wheatland Collection
Inventory No. 7571

⚙ Black and red coloring. Tablet I is a thin slice of ivory, II is a sandwich of two slices of ivory with a walnut (?) core. I is slightly warped. Horse-head brass clasp on I. Compass needle, glass, and metal ring missing.

Ia Standard German lunar volvelle with a stamped brass rotating disc (2.1 cm) with index. Scales labeled 1–12 twice (inner and middle) and 1–29 (outer). Man-in-the-moon motif on volvelle. Foliate decoration.

Ib Vertical dial, labeled VI–XII–VI in black numerals. Sun-face motif. Four small decorative stamps.

IIa Horizontal dial with a single scale for approximate latitude 47°, labeled 4–12–8 in black numerals. Compass bowl with crown maker's mark. Cardinal directions given in compass bowl as SEPT, ORI, MERI, and OCCI. Magnetic declination 5° east of north. Three small decorative stamps.

IIb Plain, except for the initials "NM" with the maker's mark (crown) between the initials. Numbered 59 in red ink. Image inverted for legibility.

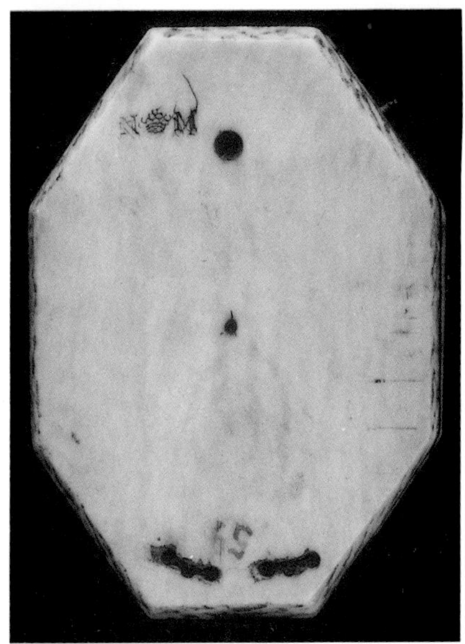

Ia | Ib
IIb | IIa

25 Octagonal Ivory and Wood Diptych

signed "NM" [Nicolaus Miller] and
maker's mark
c. 1650
Nuremberg, Germany
4.2 (w) x 5.6 (l) x 1.2 (h) cm
Formerly Harold Gillingham Collection
No. 105, then David P. Wheatland
Collection
Inventory No. 7561

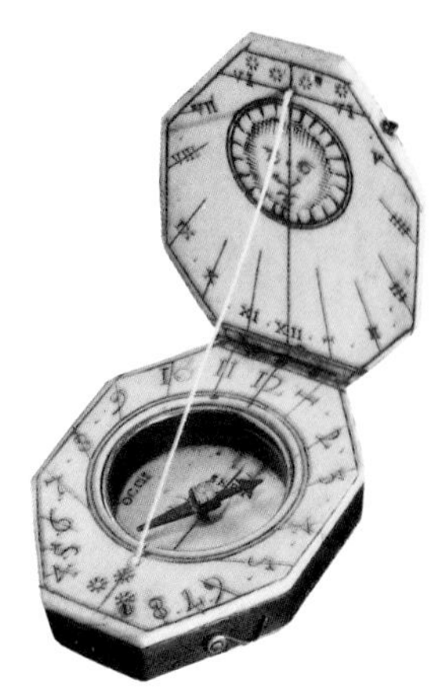

⚙ Black and red coloring. Tablet I is a thin slice of ivory, II is a sandwich of two slices of ivory with a walnut or pearwood (?) core. I is slightly warped. Horse-head brass clasp on I, wire clasps on II. Compass needle, glass, and metal ring intact.

Ia Standard German lunar volvelle with a stamped brass rotating disc (2.1 cm) with index. Scales labeled 1–12 twice (inner and middle) and 1–29 (outer). Man-in-the-moon motif on volvelle. Foliate decoration.

Ib Vertical dial, labeled VI–XII–VI in black numerals. Sun-face motif. Four small decorative stamps.

IIa Horizontal dial with a single scale for approximate latitude 46°, labeled 4–12–8 in black numerals. Compass bowl with crown maker's mark. Cardinal directions given in compass bowl as SEPT, ORI, MERI, and OCCI. Magnetic declination 4° east of north. Three small decorative stamps.

IIb Plain, except for the initials "NM" with the maker's mark (crown) between the initials. Modern paper label reads "Rorenau. Paris. 1925. 450.f." Image inverted for legibility.

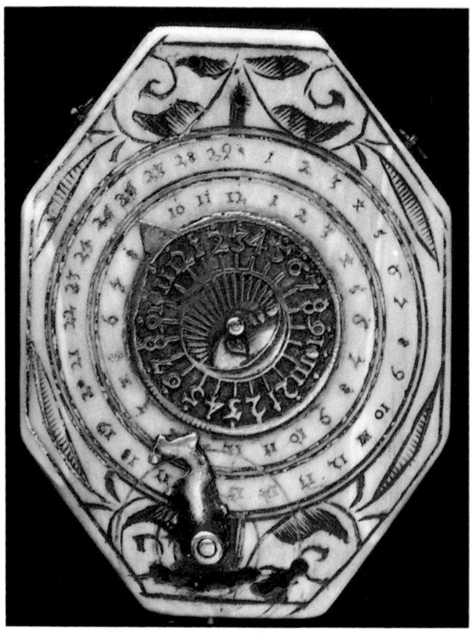

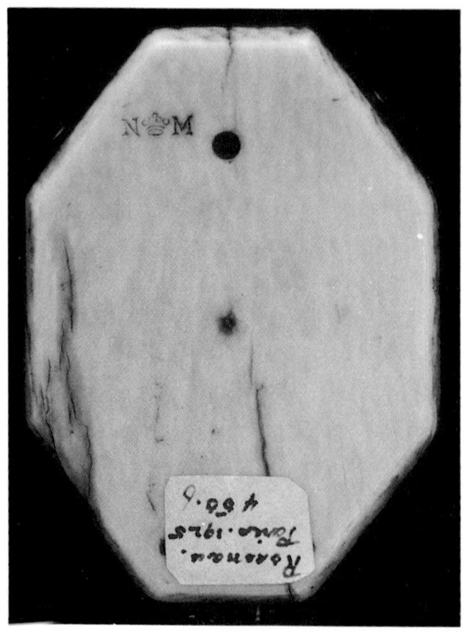

The Karner Family Workshop

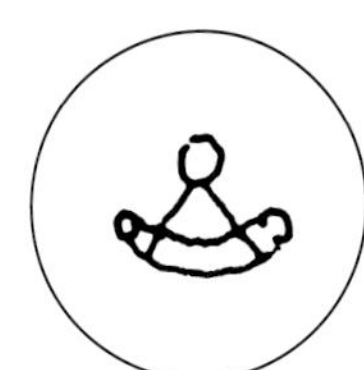

Karner family maker's marks:

Conrad

There is a continuous tradition of identifiable Karner diptychs for at least 130 years. The decoration on most Karner instruments consists mainly of punched motifs such as the snowflake, the quatrefoil, and the eight-pointed star. Like the Millers and other earlier Nuremberg *Kompassmacher* families, the Karners typically used lines, dots, and stars to indicate the half- and quarter-hours on horizontal and vertical dials.

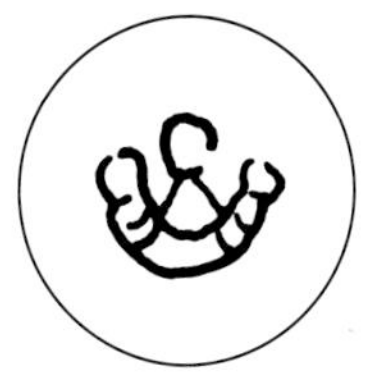

Albrecht

The earliest family member represented in this catalogue is Conrad Karner (1571–1632), who manufactured dated diptychs between 1617 and 1630 (cat. nos. 26–30 and 37). His hunting horn maker's mark was modified slightly by his son Albrecht (1619–1687, cat. nos. 33 and 39), dated examples of whose work span the years 1655 to 1668. Albrecht's older brother Jacob (born 1612) used the maker's mark "3" or, using his initials, "I3K" (cat. nos. 31 and 32; note that the letters *I* and *J* are equivalent in early modern German). Dials dated from 1636 to 1648 are attributed to Jakob. The number 3 may indicate that he was a third-generation *Kompassmacher* in his family. The fact that Albrecht's son

Jacob

Melchior (1642–1707) used the number 4 as his mark lends credence to this assumption (see Gouk, 1988, p. 117). Albrecht's younger son Georg (born 1648, cat. nos. 34–36) used a rearing horse as his mark and produced dials toward the end of the seventeenth century. The last generation of Karners represented in this collection includes Georg's son, Leonhard Andreas (born 1682, cat. nos. 43–45), whose mark was a hand. Leonhard appears to have made dials in both ivory and wood during the

Georg

1730s. A further as-yet unidentified Karner with the maker's mark "B" made both composite ivory-and-wood diptychs (cat. nos. 41 and 42) as well as wooden diptychs. Other wooden diptychs in the Harvard Collection (not catalogued here) apparently from the Karner workshop bear a tall crown mark and stem from the mid-1700s. From an analysis of the known output of the Karner workshop, there seems to be a trend in materials from ivory (seventeenth century) to composite ivory-and-wood (most common in the early eighteenth century) and finally to all-wood dials (by the mid-eighteenth century).

Leonhard Andreas?

Three of the Karner diptychs merit brief discussion. Albrecht's dial (cat. no. 33) bears a striking resemblance to the Miller family's smaller model. Perhaps Albrecht apprenticed with the Miller family, or just as likely he simply copied what appeared to be a very marketable style. Both Albrecht and Joseph Ducher frequently copied the successful Miller-style dials. Cat. no. 35 has a multitude of holes drilled into the upper tablet, which allows the string gnomon to be adjusted to a great variety of latitudes. Perhaps this diptych was owned by someone who traveled extensively. Cat. no. 34 is exceptional in that it includes one of the only three

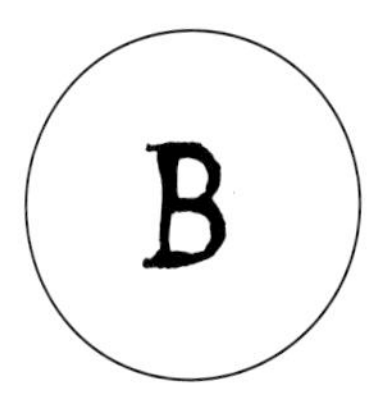

?

Nuremberg wind vanes known to the author to be still extant (see figure 6, page 29). Such auxiliary parts were often lost over the centuries.

One of the highlights of Harvard's collection is the multitude of miniature diptychs manufactured by the Karner workshop (cat. nos. 37–45). Of these nine, seven are tiny composite ivory-and-wood diptychs. One of these is complete with its original black leather case (cat. no. 45). A

Michael Lesel

curious feature of several of these miniature diptychs is the presence of a volvelle scale simply counting from 1 to 12 (cat. nos. 40 and 42); such a scale cannot serve as a lunar volvelle to convert to local apparent solar time at night. The use (if any) of the scale remains a mystery, even more so because the accompanying brass volvelles are missing. One might suggest that these dials were originally equipped with tidal volvelles, which became popular on the French ivory dials of the mid- to later-seventeenth century; however, there was no apparent need for tidal volvelles in German cities, most of which were landlocked. At least two of the miniature diptychs are known to be made from bone rather than ivory (one in the Lewis Evans Collection at Oxford, the other our cat. no. 38).

Cat. no. 46 bears the crown maker's mark used both by Paul Reinmann and by Michael Lesel, but the use of bright colors and other stylistic elements (such as flowers, fruits, and putti-faced wind motifs) lead one to assign this diptych to Michael Lesel. Lesel's relation by marriage to the Karner family explains his inclusion in this catalogue with the Karner workshop. His father Albrecht, also a *Kompassmacher,* married Elisabeth, the widow of the dial-maker Caspar Karner (see Gouk, 1988, p. 53–55).

26 Oval Ivory Diptych

signed "CONRAD KARNER" and
maker's mark
dated 1620
Nuremberg, Germany
5.1 (w) x 7.0 (l) x 1.3 (h) cm
Formerly Harold Gillingham Collection
No. 103, then David P. Wheatland
Collection
Inventory No. 7542

⬤ Faded black, brown, and green coloring. Brass wire clasps on I and II, hanging
loop on II. Compass needle and brass ring
intact, glass missing. Hole in II for attaching
string gnomon. String gnomon is attached
to a brass plumb bob to hold it taut.

Ia Undecorated.

Ib Vertical dial labeled VI–XII–VI in faded
black numerals. Large sun-face motif.
Signed "CONRAD KARNER" under sun-
face. Dated 1620 at top of dial.

IIa Horizontal dial with a single scale for
approximate latitude 48°, labeled 4–12–8
in black numerals. Maker's mark (a hunting
horn) appears twice in compass bowl.
Cardinal directions given in compass bowl
as SEPT, ORIE, MERI, and OCCI. Magnetic
declination 4° east of north. Green and
brown foliate decoration.

IIb Undecorated, except for the maker's
mark above string gnomon hole. Modern
paper label reads "Conrad Karner. German
1620/45 Hof. Geneva 250. f." Image
inverted for legibility.

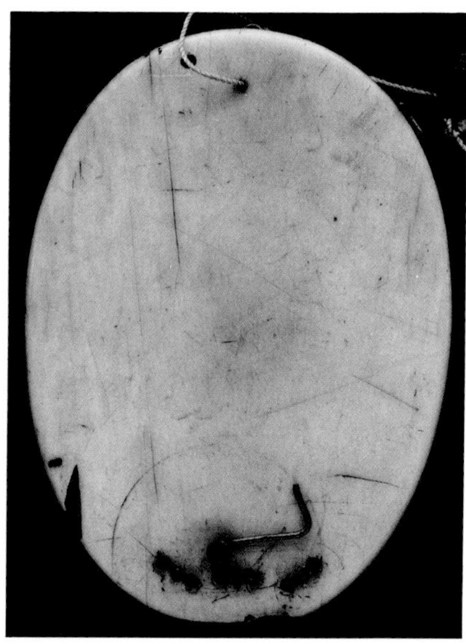

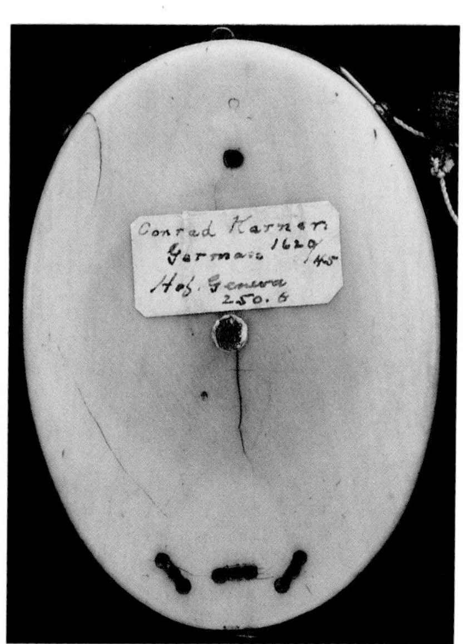

<table>
<tr><td>Ia</td><td>Ib</td></tr>
<tr><td>IIb</td><td>IIa</td></tr>
</table>

27 Octagonal Ivory Diptych

signed "CONRAD KARNER" and
maker's mark
c. 1620
Nuremberg, Germany
5.2 (w) x 7.2 (l) x 1.4 (h) cm
Formerly David P. Wheatland Collection
(purchased in France, 1952)
Inventory No. 7539

⬤ Black, red, and blue coloring. Brass
wire clasps on I and II, hanging loop on II.
Compass needle intact, glass and metal ring
missing. Hole in II for attaching string.

Ia Standard German lunar volvelle with a
stamped brass rotating disc (2.5 cm) with
index. Scales labeled 1–12 twice (inner and
middle) and 1–29 (outer). Man-in-the-
moon motif on volvelle. Foliate decoration.

Ib Vertical dial, scale labeled VI–XII–VI in
black numerals. Large eight-pointed star
motif with sun stamps. Signed "CONRAD
KARNER" under star. Four small decorative
stamps.

IIa Horizontal dial with a single scale for
approximate latitude 48.5°, labeled 4–12–8
in black numerals. Maker's mark (a hunting
horn) appears twice in compass bowl. Cardinal directions given in compass bowl as
SEPT, ORIE, MERI, and OCCI. Magnetic declination 4° east of north. Pin gnomon dial
for Italian hours labeled 10–23 in black
numerals. Three small decorative stamps.

IIb Plain, except for the maker's mark
(hunting horn) next to string gnomon
hole.

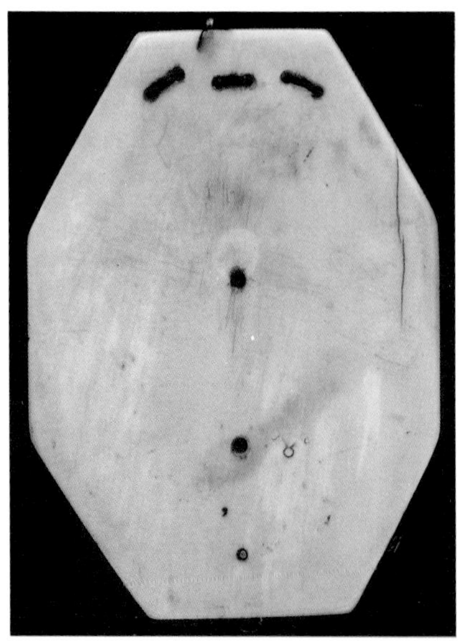

Ia | Ib
IIb | IIa

28 Rectangular Ivory Diptych

signed "CONRAD KARNER" and
maker's mark
dated 1622
Nuremberg, Germany
7.6 (w) x 11.6 (l) x 1.6 (h) cm
Formerly Drecker Collection No. 131,
then David P. Wheatland Collection
Inventory No. 7552

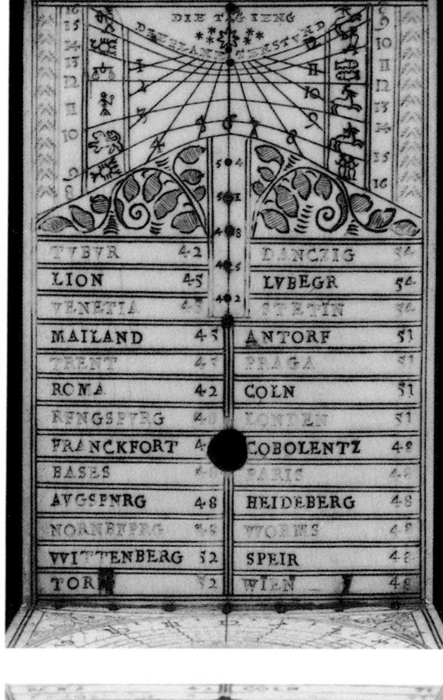

⊙ Faded black, red, and green coloring.
Tablet II is beveled lengthwise along edges.
Compass viewing-hole in I, slot for wind
vane in side of II. Brass plate clasp on I,
hanging loop, one bun foot on I and four
feet on II. Compass needle intact, glass and
metal ring missing.

Ia Wind rose with 32 points, numbered
1–32 beginning at east, 16 directions
labeled or abbreviated in German, wind
vane missing. Foliate decoration.

Ib Pin gnomon dial (gnomon missing)
labeled "DIE TAG IENG" [length of the day]
and 16–8 and 8–16 in red numerals, labeled
"DIE PLANETEN STVND" [planetary hours]
and 1–12 in black numerals, and marked
with zodiacal symbols. Holes for string
gnomon labeled 42°, 45°, 48°, 51°, and
54°. Latitude table (42°–54°) for 26
European cities, marked alternately in red
and black. Green leaves.

IIa Horizontal dial with five scales for
latitudes 42° (inner), 45°, 48°, 51°, and 54°
(outer), labeled 4–12–8 in black numerals
in outermost scale. Pin gnomon dial for
Italian hours (labeled 9–23 in red numerals)
and Babylonian hours (labeled 1–15 in
black numerals). Maker's mark (hunting
horn) appears in compass bowl, partially
worn away. Hole at bottom of compass
bowl. Surface of compass bowl badly
defaced: only "OCCI" is readable. Magnetic
declination 5° east of north. Green and
brown foliate decoration. Signed "CONRAD
KARNER" at bottom.

IIb Standard German lunar volvelle with a
stamped brass rotating disc (2.4 cm) with
index. Scales labeled 1–12 twice (inner and
middle) and 1–29 (outer). Man-in-the-
moon motif on volvelle. Epact scales
(beginning with 23 and 13, respectively, for
the year 1608) around volvelle, labeled
"EPACTA IVLIANI ANNO 1622" [Julian epacts,
year 1622] and "EPACTA CRECORI ANNO"
[Gregorian epacts, year]. Foliate decoration. Numbered 168 twice near hinge.

29 Rectangular Ivory Diptych

signed "CONRAD KARNER" and
maker's mark
dated 1630
Nuremberg, Germany
6.3 (w) x 9.2 (l) x 1.3 (h) cm
Formerly Drecker Collection No. 132,
then David P. Wheatland Collection
Inventory No. 7553

⊗ Faded black, red, and green coloring.
Compass viewing-hole in I, slot with plain
brass clasp for wind vane in side of II. Brass
hook clasps on I and II, hanging loop, three
bun feet on I and four on II. Compass
needle and glass intact, metal ring missing.

Ia Wind rose with 32 points, 16 labeled or
abbreviated in German, numbered 1–32
beginning at east, wind vane missing.
Foliate decoration.

Ib Pin gnomon dial labeled 16–8 and 8–16
in red numerals to indicate hours of daylight and darkness, hour lines labeled 1–12
for planetary hours, and marked with
zodiacal symbols. Holes for string gnomon
labeled 42°, 45°, 48°, and 51°. Latitude
table for 18 European cities (42°–51° latitude). Green leaves and several decorative
stamps.

IIa Horizontal dial with four scales for
latitudes 42° (inner), 45°, 48°, and 51°
(outer), labeled 4–12–8 in black numerals
in outermost scale. Separate pin gnomon
dials for Italian hours (left, labeled 10–23
in black numerals) and Babylonian hours
(right, labeled 1–14 in black numerals).
Maker's mark (hunting horn) appears twice
in compass bowl. Cardinal directions given
in compass bowl as SEPT, ORIE, MERI, and
OCCI. Magnetic declination 3° east of
north. Green leaves. Signed "CONRAD
KARNER" at bottom.

IIb Standard German lunar volvelle, brass
disc missing. Scales labeled 1–12 twice
(inner) and 1–29 (outer). Epact scale
(beginning with 23 and 13, respectively, for
the year 1627) in circular form around volvelle, labeled "EPACTA IVLIANI ANNO 1630"
[Julian epacts, year 1630] and "EPACTA
GREGORI ANNO" [Gregorian epacts, year].
Maker's mark (hunting horn) appears again

near hinge. Numbered 7 twice near hinge
in red ink (once on a modern paper label).
Foliate decoration.

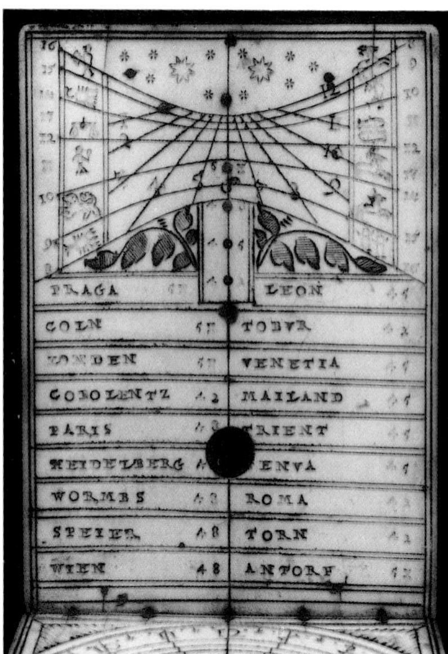

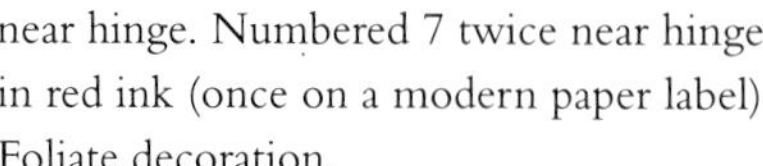

Ia | Ib
IIb | IIa

30 Rectangular Ivory Diptych

signed "CONRAD KANNER" and
maker's mark
dated 1630
Nuremberg, Germany
5.3 (w) x 8.2 (l) x 1.1 (h) cm
Formerly Drecker Collection No. 133,
then David P. Wheatland Collection
Inventory No. 7554

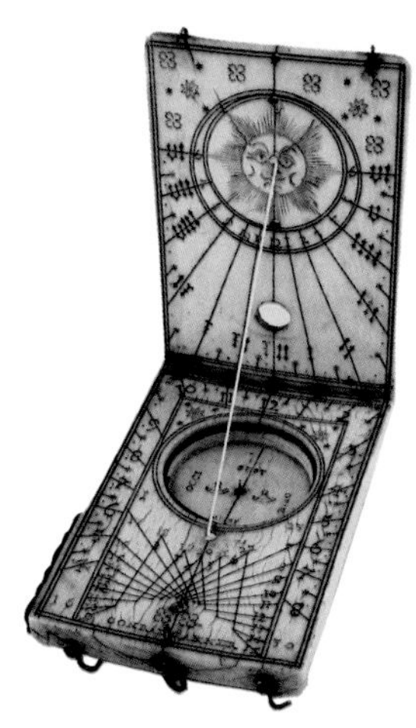

⚙ Black and red coloring. Both tablets
slightly warped (curved). Compass
viewing-hole in I, slot with brass clasp for
wind vane in side of II. Brass wire clasps
on I and II, hanging loop and four bun feet
on II, one foot on I. Compass needle and
glass missing. Metal ring intact.

Ia Wind rose with 32 points labeled or
abbreviated in German, numbered 1–32
beginning at east, index and wind vane
missing. Foliate decoration.

Ib Vertical dial labeled VI–XII–VI in black
numerals and 6–12–6 in red numerals.
Decorated with a sun-face and several
decorative stamps.

IIa Horizontal dial with a single scale for
approximate latitude 47.5°, labeled 4–12–8
in black numerals. Pin gnomon dial for
Italian hours (labeled 9–23 in red numerals)
and Babylonian hours (labeled 1–15 in
black numerals). Compass bowl with
maker's mark twice (hunting horn). Cardinal directions given in compass bowl as
SEPT, ORIE, MERI, and OCCI. Magnetic
declination 4° east of north. Several decorative stamps. Signed "CONRAD KANNER"
and dated 1630 near bottom.

IIb Standard German lunar volvelle, brass
disc missing. Scales labeled 1–12 twice
(inner) and 1–29 (outer). Two maker's
marks (hunting horns) appear again near
hinge. Foliate decoration. Numbered 2 in
red ink near center, and again on a modern
paper label near the hinge.

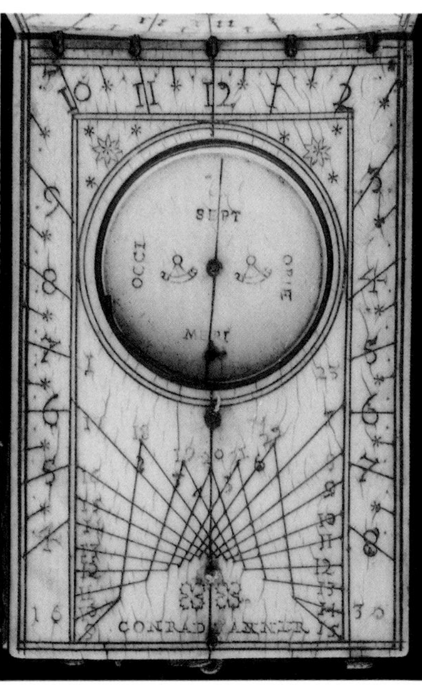

Ia	Ib
IIb	IIa

31 Rectangular Ivory Diptych

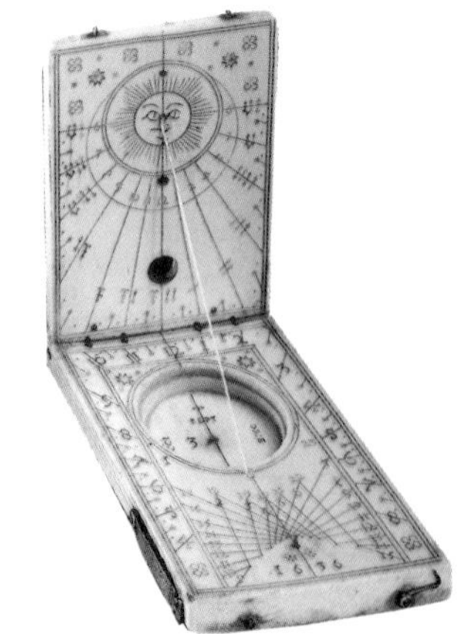

signed with maker's mark [Jacob Karner]
dated 1636
Nuremberg, Germany
5.4 (w) x 8.5 (l) x 1.3 (h) cm
Formerly Drecker Collection No. 139,
then David P. Wheatland Collection
Inventory No. 7550

⊙ Black and red coloring. Compass viewing-hole in I, slot with brass clasp for wind vane in side of II. Brass hook clasps on II, wire clasp on I, four bun feet on II. Compass needle, glass, and ring missing. Hole in II for attaching string.

Ia Wind rose with 32 compass points, numbered 1–32 beginning at east, 16 directions labeled or abbreviated in German. Brass arm-shaped indicator. Black and red foliate decoration in corners.

Ib Vertical dial labeled VI–XII–VI in black numerals and 6–12–6 in red numerals. Red asterisks indicate the half-hours. Large sun-face motif. Two black eight-pointed stars and six red quatrefoil stamps.

IIa Horizontal dial with a single scale for approximate latitude 49°, labeled 4–12–8 in black numerals. Maker's mark ("3") appears in compass bowl. Cardinal directions given in compass bowl as SEPT, ORIE, MERI, and OCCI. Magnetic declination 4° east of north. Pin gnomon dial for Italian hours (labeled 9–23 in red numerals) and Babylonian hours (labeled 1–15 in black numerals). Four small black eight-pointed stars and two red quatrefoil stamps. Dated 1636 near bottom.

IIb Standard German lunar volvelle with a stamped brass rotating disc (2.5 cm) with index. Scales labeled 1–12 twice (inner and middle) and 1–29 (outer). Man-in-the-moon motif on volvelle. Black and red foliate decoration in corners. Maker's mark "3" appears near hinge. Surface of ivory is badly yellowed across most of dial face. Numbered 94 on small modern paper label near hinge.

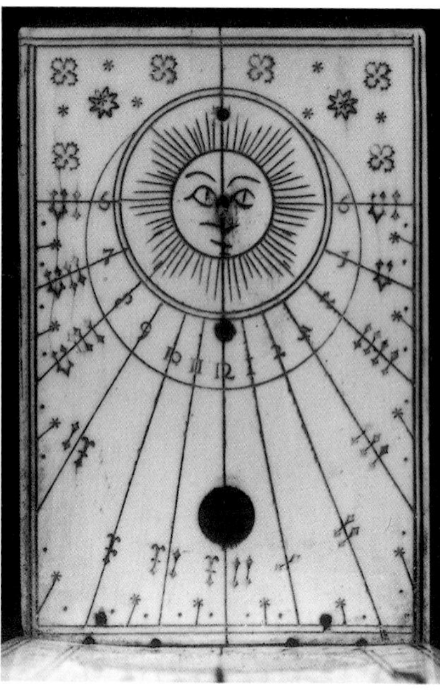

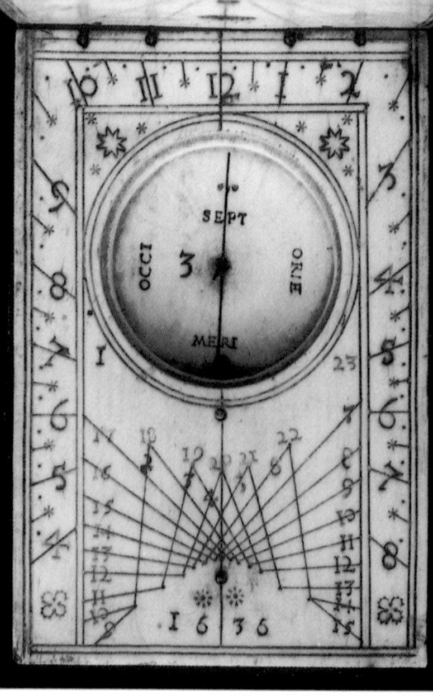

Ia | Ib
IIb | IIa

32 Rectangular Ivory Diptych

signed with maker's mark [Jacob Karner]
c. 1640
Nuremberg, Germany
4.9 (w) x 7.2 (l) x 1.3 (h) cm
Formerly Drecker Collection No. 140,
then David P. Wheatland Collection
Inventory No. 7541

⬤ Black and red coloring. Compass viewing-hole in I, slot with brass clasp for wind vane in side of II. Brass wire clasps on I and II. Compass needle and glass intact, copper ring is a replacement. Hole in II for attaching string.

Ia Standard German lunar volvelle with an engraved and punched brass rotating disc (2.5 cm) with index. Red and blue curves seen through hole in brass disc indicate lunar phase. Scales labeled 1–12 twice (inner and middle) and 1–29 (outer). Sixteen compass directions labeled in German arranged concentrically around volvelle. Decoration in corners.

Ib Vertical dial labeled VI–XII–VI in black numerals and 6–12–6 in red numerals. Red asterisks indicate the half-hours. Large six-pointed star motif.

IIa Horizontal dial with a single scale for approximate latitude 48°, labeled 4–12–8 in black numerals. Maker's mark ("3") appears in compass bowl. Cardinal directions given in compass bowl as SEPT, ORIE, MERI, and OCCI. Magnetic declination zero degrees. Pin gnomon dial for Italian hours labeled 10–23 in black numerals. Four small eight-pointed decorative stamps.

IIb Plain, except for the maker's mark ("13K") above the string gnomon hole. Numbered 146 on modern paper label in black ink and near top edge in red ink. Image inverted for legibility.

33 Rectangular Ivory Diptych

signed "ALBRECHT KARNER" and
maker's mark
dated 1655
Nuremberg, Germany
5.6 (w) x 9.5 (l) x 1.2 (h) cm
Formerly David P. Wheatland Collection
(purchased in France, 1952)
Inventory No. 7540

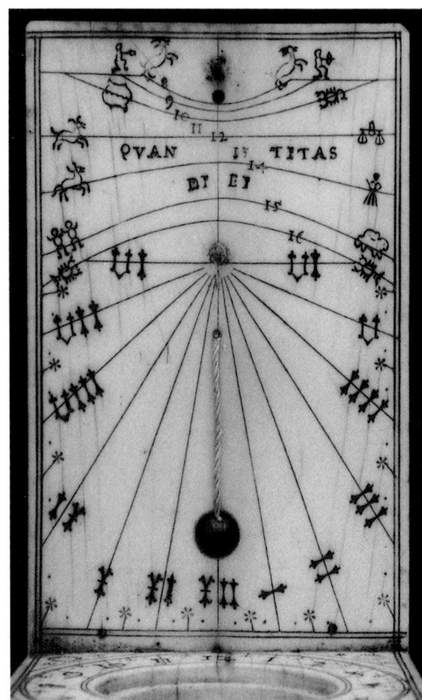

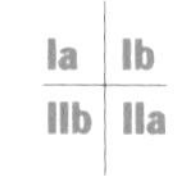

⬤ Black, red, brown, and green coloring.
Compass viewing-hole in I, slot with brass
clasp for wind vane in side of II. Brass wire
clasps on I and II and four bun feet on II.
Small brass ball is a plumb bob to level the
dial; the ball fits into the compass viewing-
hole when the diptych is closed. Compass
needle, glass, and metal ring missing.

Ia Wind rose with 16 points and 16 directions labeled in German, wind vane
missing. Sun-face motif at center of wind
rose. Brown and green foliate decoration.

Ib Pin gnomon dial labeled "QVANTITAS
DIEI" [length of the day] and 8–16 in red
numerals, and marked with zodiacal
symbols. Vertical dial labeled VI–XII–VI in
black numerals. Red asterisk stamps mark
the half-hours.

IIa Horizontal dial with a single scale for
approximate latitude 49°, labeled 5–12–7
in black numerals. Pin gnomon dial for
Babylonian hours (labeled 1–14 in red
numerals) and Italian hours (labeled 10–23
in black numerals). Compass bowl with
four-winds motif , cardinal directions given
as SE, OR, ME, and OC, and maker's mark
(hunting horn). Magnetic declination
indicated at 5° east and 16° west of north.
Two small sun decorations. Small red asterisk stamps. Signed "ALBRECHT KARNER
1655" near bottom.

IIb Standard German lunar volvelle with a
stamped brass rotating disc (2.1 cm) with
index. Scales labeled 1–12 twice (inner and
middle) and 1–30 (outer). Man-in-the-
moon motif on volvelle. Outer two scales
are for correction between lunar and solar
time (in hours and minutes). Reddish-
brown and green foliate decoration.
Maker's mark (hunting horn) appears twice
near hinge. Image inverted for legibility.

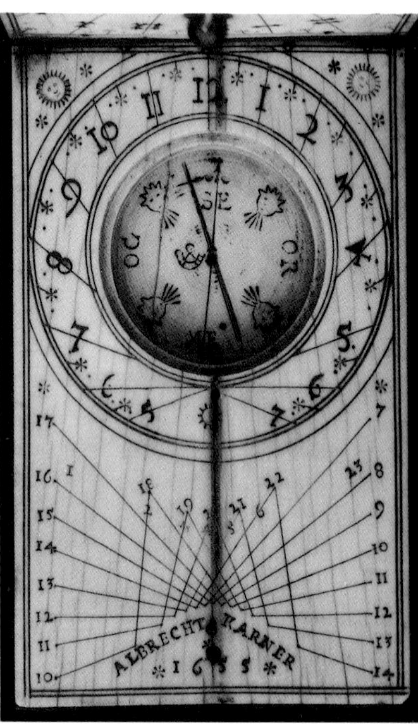

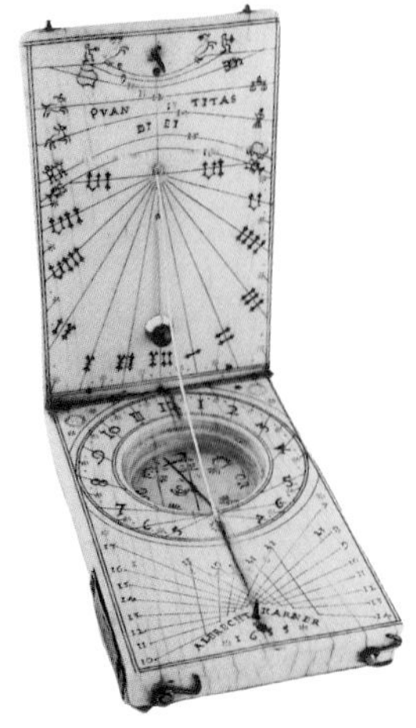

34 Rectangular Ivory Diptych

signed with maker's mark [Georg Karner]
dated 1700
Nuremberg, Germany
6.2 (w) x 10.8 (l) x 1.4 (h) cm
Formerly Harold Gillingham Collection
No. 88, then David P. Wheatland
Collection
Inventory No. 7525

⊕ Black, red, brown, and green coloring.
Leaf I slightly warped. Compass viewing-
hole in I, slot with brass clasp for wind
vane in side of II. Brass scythe clasps on I
and II, four bun feet on II. Compass
needle, glass, and braided wire ring intact.

Ia Wind rose with 32 points, numbered
counter-clockwise 1–32 beginning at
north, 32 directions labeled in German.
Brass arrow index and wind vane. (See
figure 6, page 29.) Green and brown foliate
decoration.

Ib Pin gnomon labeled "QVAWTITAS DIRI"
[length of the day] and 8–16 in red
numerals, and marked with zodiacal symbols. Holes for string gnomon at 42°, 45°,
48°, 51°, and 54°. Latitude table for 20
European cities (42°–55° latitude). Sun-
face motif. Engraving of two castles amidst
mountains.

IIa Horizontal dial with four scales for
latitudes 42° (inner), 45°, 48°, and 51°
(outer), labeled 4–12–8 in black numerals.
Compass bowl with four-winds motif and
rearing horse maker's mark twice. Cardinal
directions given in compass bowl as SEPT,
ORIE, MERI, and OCCI. Magnetic declination 5° west of north. Pin gnomon dial for
Italian hours (labeled 10–23 in black
numerals) and Babylonian hours (labeled
1–14 in red numerals). Sun-face motif.
Foliate decoration.

IIb Standard German lunar volvelle with a
stamped brass rotating disc (2.6 cm) with
index. Scales labeled 1–12 twice (inner and
middle) and 1–29 (outer). Epact scale
(beginning with 23 and 13, respectively, for
the year 1684) in circular form around volvelle, labeled "EPACTA IVLIANI ANNO 1700"
[Julian epacts, year 1700] and "EPACTA
GREGORI ANNO" [Gregorian epacts, year].
Green and brown foliate decoration.

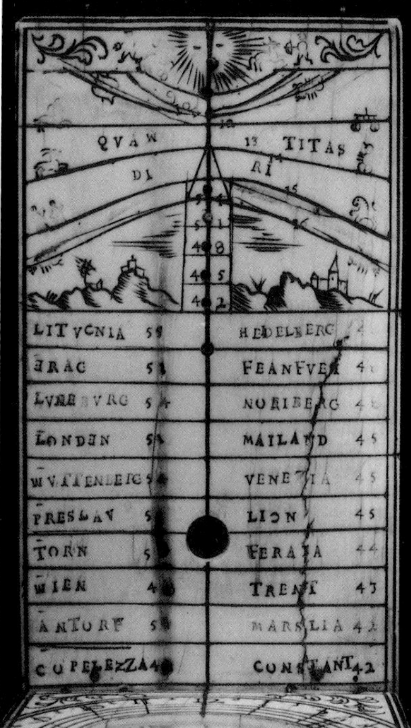

Ia | Ib
IIb | IIa

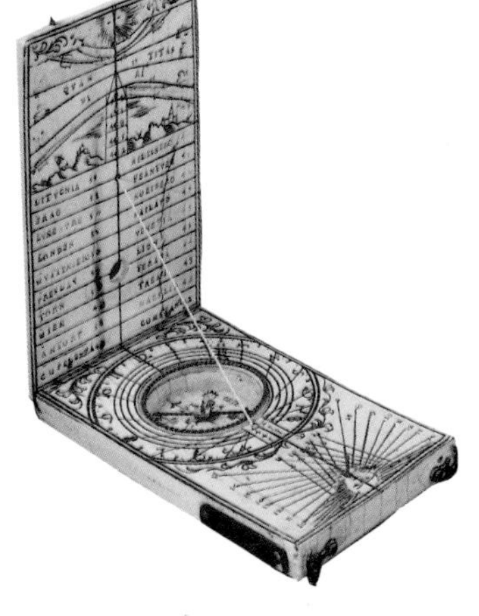

35 Rectangular Ivory Diptych

unsigned [Georg (?) Karner]
c. 1700
Nuremberg, Germany
7.2 (w) x 9.8 (l) x 1.4 (h) cm
Formerly Harold Gillingham Collection
No. 87, then David P. Wheatland
Collection
Inventory No. 7524

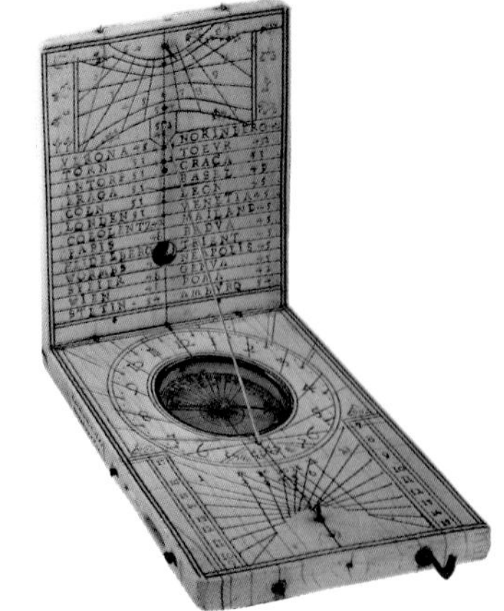

⊚ Black and red coloring. Tablet I slightly warped. Compass viewing-hole in I, slot for wind vane in side of II. Brass wire clasps on I and II, a single bun foot each on I and II. Compass needle, glass, and metal ring intact. Modern paper label along left outer edge reads "German. c. 1650. Mercator. 1500."

Ia Wind rose with 32 points, labeled 1–32 (west = 32), 16 directions labeled or abbreviated in German. Wind vane missing. Double-lined border.

Ib Pin gnomon dial with hour lines labeled 1–12 in red numerals for planetary hours, and marked with zodiacal symbols. Holes for string gnomon labeled at 42°, 45°, 48°, 51°, and 54°, with four additional unlabeled holes for lower latitudes (probably added at a later date). Latitude table for 26 European cities (42°–54° latitude). Double-lined border.

IIa Horizontal dial with a single scale labeled 5–12–7 in black numerals. Pin gnomon dial for Italian hours (labeled 9–23 in red numerals) and Babylonian hours (labeled 1–15 in black numerals). Compass bowl with painted paper insert graduated to 50° east or west of north to account for changes in magnetic declination (added later). Quatrefoil and other decorative stamps. Double-lined border.

IIb Standard German lunar volvelle, brass disc missing. Scales labeled 1–12 twice (inner) and 1–29 (outer). Epact scale (beginning with 23 and 13, respectively, for the year 1684) in circular form around volvelle, labeled "EPACTA IVLIANI ANNO" [Julian epacts, year] and "EPACTA GREGORI ANNO" [Gregorian epacts, year]. Double-lined border.

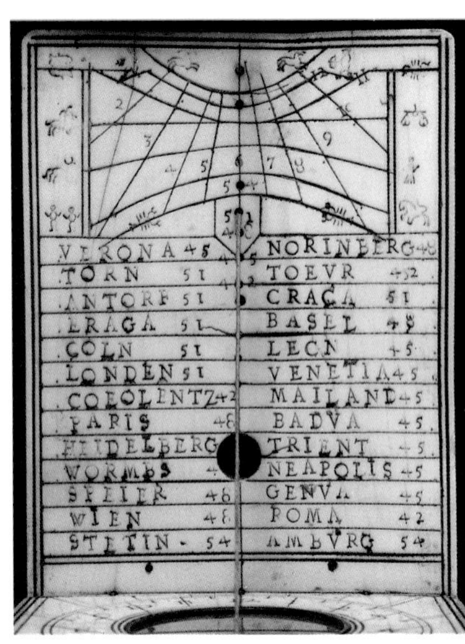

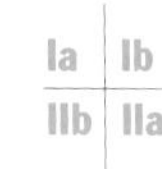

Ia | Ib
IIb | IIa

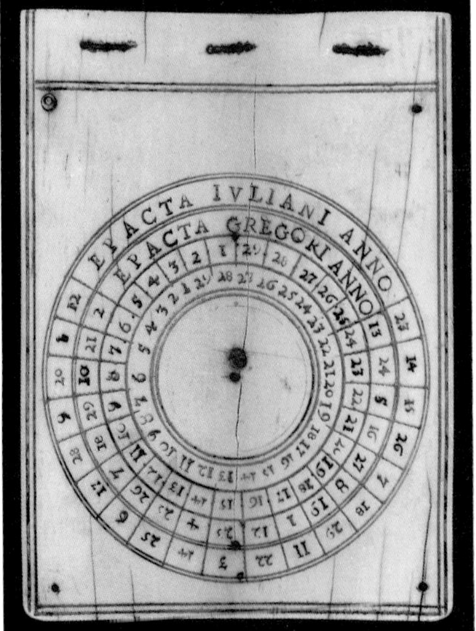

36 Rectangular Ivory Diptych

signed with maker's mark [Georg Karner]
c. 1700
Nuremberg, Germany
6.1 (w) x 9.4 (l) x 1.3 (h) cm
Formerly David P. Wheatland Collection
(purchased in France, 1952)
Inventory No. 7526

⚙ Black and red coloring. Tablet I slightly warped. Compass viewing-hole in I, slot for wind vane in side of II. Brass wire clasps on I and II, four bun feet on II. Compass needle intact, glass and metal ring missing.

Ia Wind rose with 16 points, 16 directions labeled in German, wind vane missing.

Ib Vertical dial labeled VI–XII–VI in black numerals. Decorated with large six-pointed star and alternating red and black circular motifs.

IIa Horizontal dial with a single scale for approximate latitude 48°, labeled 5–12–7 in black numerals. Compass bowl with four-winds motif and rearing horse maker's mark. Cardinal directions given in compass bowl as SEPT, ORIE, MERI, and OCCI. Magnetic declination 10° west of north. Pin gnomon dial for Italian hours (labeled 10–23 in red numerals) and Babylonian hours (labeled 1–14 in black numerals). Foliate and circular red decorative stamps.

IIb Standard German lunar volvelle with a stamped brass rotating disc (2.5 cm) with index. Scales labeled 1–12 twice (inner and middle) and 1–29 (outer). Man-in-the-moon motif on volvelle.

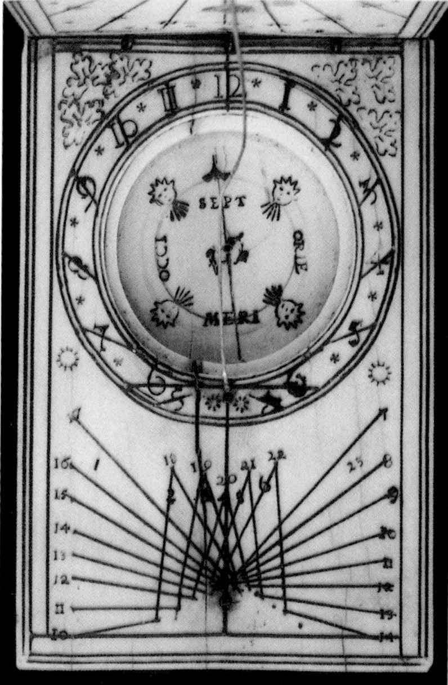

Ia	Ib
IIb	IIa

37 Miniature Ivory and Wood Diptych

signed with maker's mark
[Conrad Karner]
c. 1620
Nuremberg, Germany
3.1 (w) x 4.5 (l) x 0.9 (h) cm
Formerly Drecker Collection
No. 135, then David P.
Wheatland Collection
Inventory No. 7546

38 Miniature Bone Diptych

unsigned [Karner workshop?]
17th or early 18th century
Nuremberg, Germany
2.7 (w) x 4.5 (l) x 0.9 (h) cm
Formerly Ernst Collection
No. 94
Inventory No. 7894

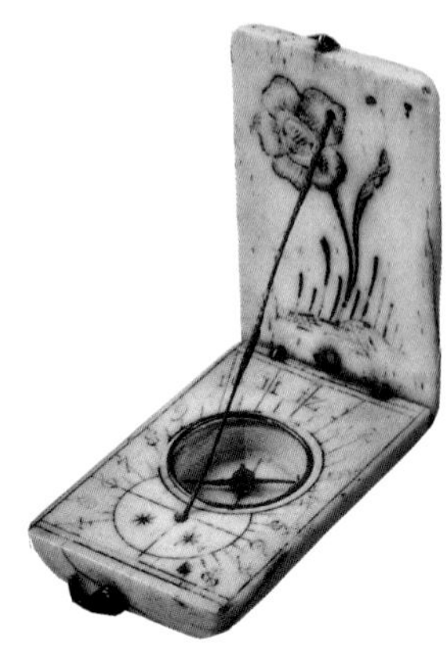

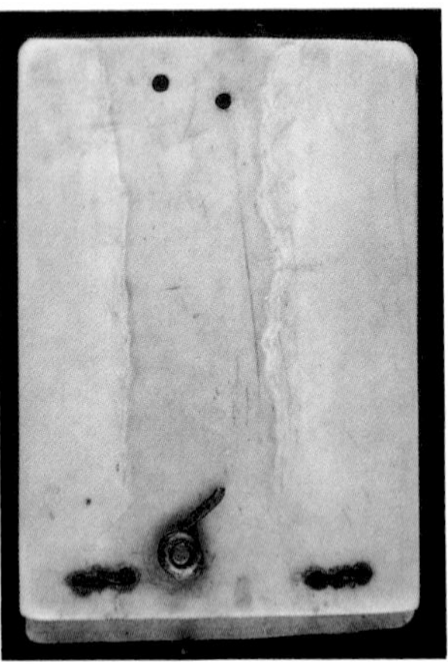

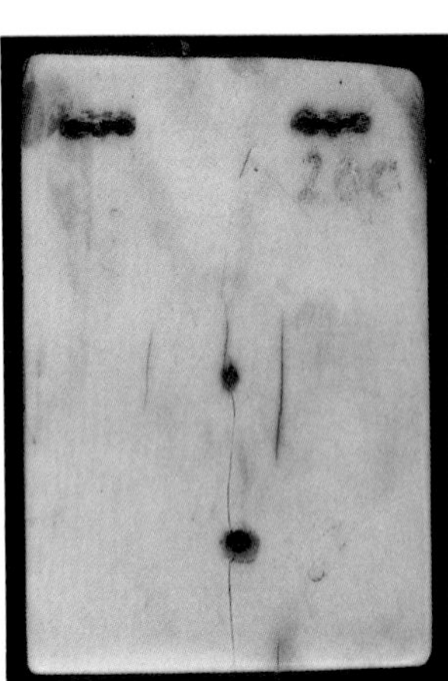

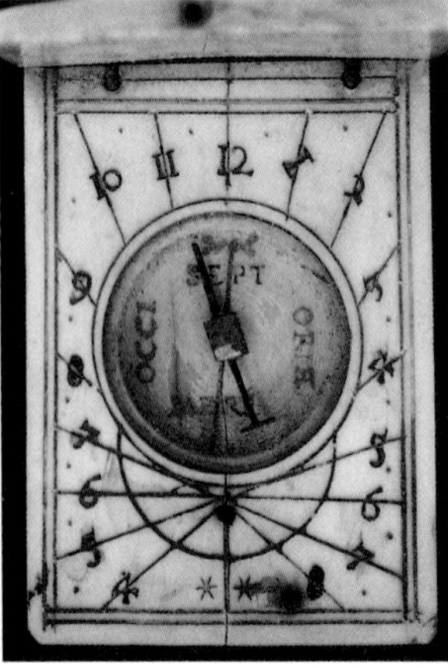

● Faded black, red, and blue coloring. Outer face of I is curved. Brass wire clasps on I broken. Brass pin on face IIa to keep dial leaves aligned. Compass needle and glass intact. Hole in II through which to attach string. II is cracked lengthwise between this attachment hole and the edge.

Ia Undecorated.

Ib Vertical dial labeled VI–XII–VI in black numerals. Circular decoration.

IIa Horizontal dial with a single scale for approximate latitude 48°, labeled 4–12–8 in black numerals. Cardinal directions given in compass bowl. Magnetic declination about 8° east of north. Two small red star stamps.

IIb Undecorated, except for a very faint maker's mark (hunting horn) near string gnomon hole. It resembles the mark of Conrad Karner more than that of his son, Albrecht. Numbered 206(?) in red ink.

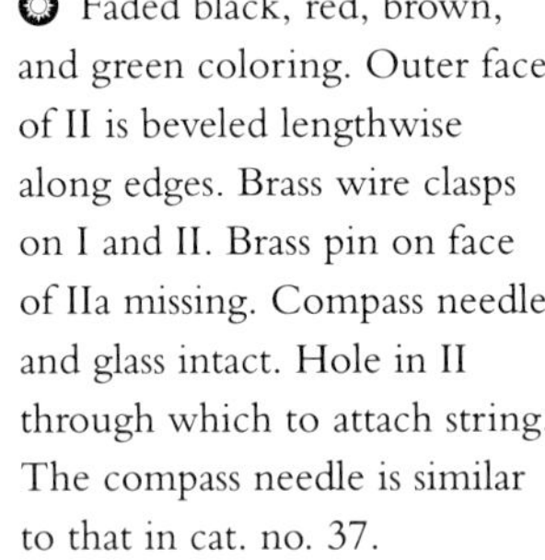

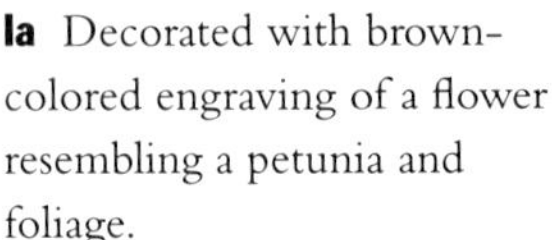

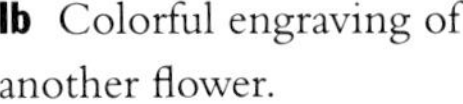

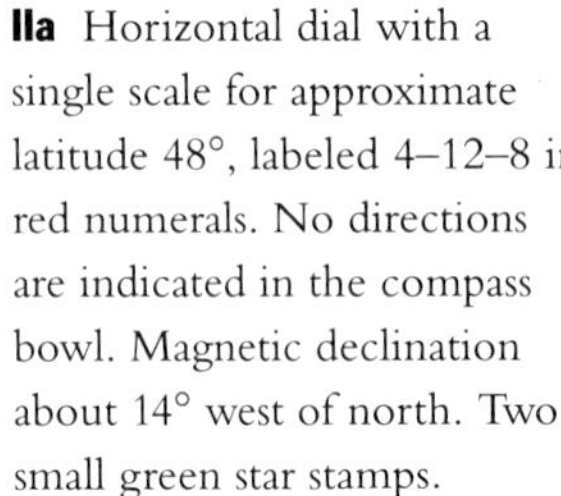

● Faded black, red, brown, and green coloring. Outer face of II is beveled lengthwise along edges. Brass wire clasps on I and II. Brass pin on face of IIa missing. Compass needle and glass intact. Hole in II through which to attach string. The compass needle is similar to that in cat. no. 37.

Ia Decorated with brown-colored engraving of a flower resembling a petunia and foliage.

Ib Colorful engraving of another flower.

IIa Horizontal dial with a single scale for approximate latitude 48°, labeled 4–12–8 in red numerals. No directions are indicated in the compass bowl. Magnetic declination about 14° west of north. Two small green star stamps.

IIb Undecorated. Numbered 94 in black ink.

39 Miniature Ivory and Wood Diptych

signed with maker's mark
[Albrecht Karner]
mid-17th century
Nuremberg, Germany
3.2 (w) x 4.6 (l) x 1.1 (h) cm
Formerly Drecker Collection
No. 134, then David P.
Wheatland Collection
Inventory No. 7543

☀ Faded black, red, and blue coloring. Tablet I is a thin slice of ivory, II is a sandwich of two slices of ivory with a walnut or pearwood (?) core. Brass wire clasps on I and II. Brass pin on face of IIa to keep the two leaves aligned when the dial is closed. Compass needle missing and glass intact. Hole in II through which to attach string.

Ia Standard German lunar volvelle with a stamped brass rotating disc (1.8 cm) with index. Scales labeled 1–12 twice (inner and middle) and 1–29 (outer). Man-in-the-moon motif on volvelle. Decoration in corners.

Ib Vertical dial labeled VI–XII–VI in faint black numerals. Two small stamps.

IIa Horizontal dial with a single scale for approximate latitude 48°, labeled 4–12–8 in black numerals. Magnetic declination about 5° east of north. Maker's mark (hunting horn) in compass bowl. Cardinal directions given in compass bowl as SE, OR, ME, and OC. Three small decorative stamps.

IIb Undecorated, except for maker's mark (hunting horn) near string gnomon hole. Numbered 42 in red ink near hinge and in red ink on a small modern paper label near center. Image inverted for legibility.

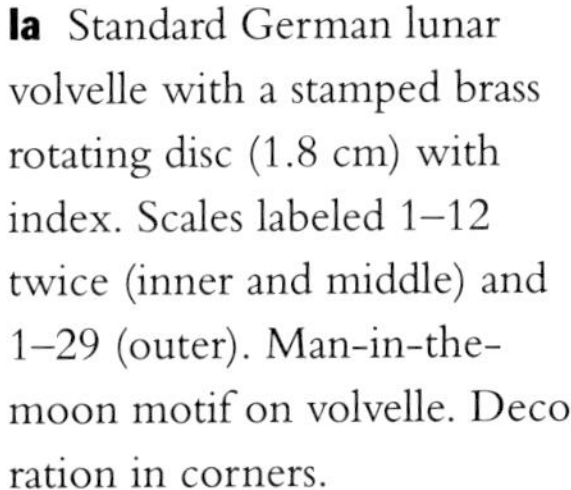
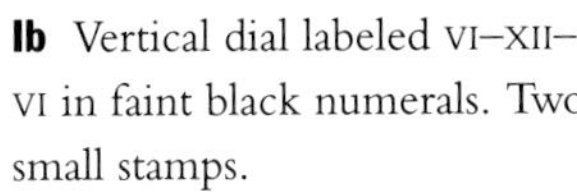

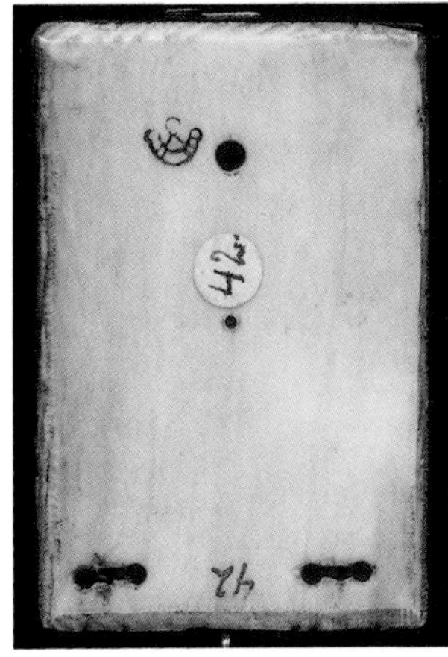

Ib
IIa

Ia Ib
IIb IIa

40 Miniature Ivory and Wood Diptych

unsigned [Karner workshop]
first half of 18th century
Nuremberg, Germany
2.8 (w) x 4.3 (l) x 1.3 (h) cm
Formerly Harold Gillingham
Collection No. 106, then
David P. Wheatland Collection
Inventory No. 7521

41 Miniature Ivory and Wood Diptych

signed with maker's mark
[Karner workshop]
first half of 18th century
Nuremberg, Germany
3.2 (w) x 4.8 (l) x 1.2 (h) cm
Formerly David P. Wheatland
Collection (purchased in
France, 1952)
Inventory No. 7544

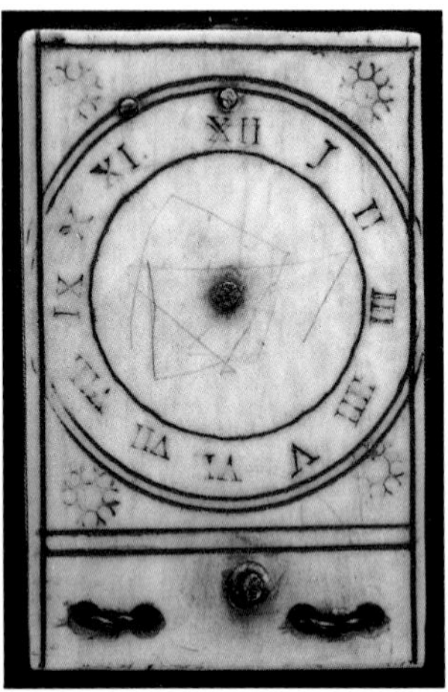

la | lb
llb | lla

⚙ Black, red, and blue color-ing. Tablet I is a thin slice of ivory, II is a sandwich of ivory with a walnut or pearwood (?) core. Brass wire clasp on II. Brass pin on face of IIa to keep the two leaves aligned when the dial is closed. Glass and metal ring intact, compass needle replaced (?). Hole in II for attaching string.

la Volvelle with only one scale from I to XII, rotating disc miss-ing. Decoration in corners.

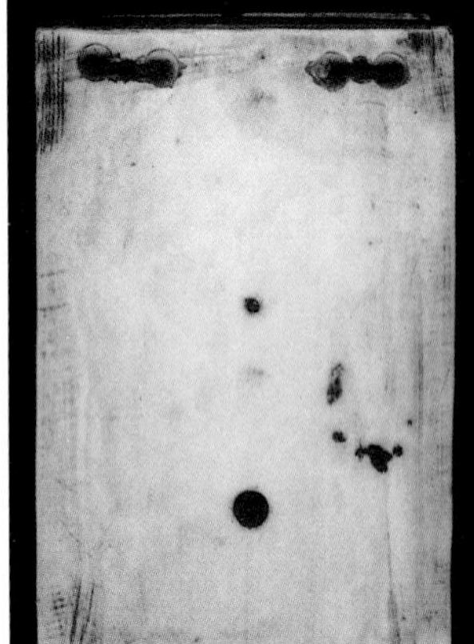

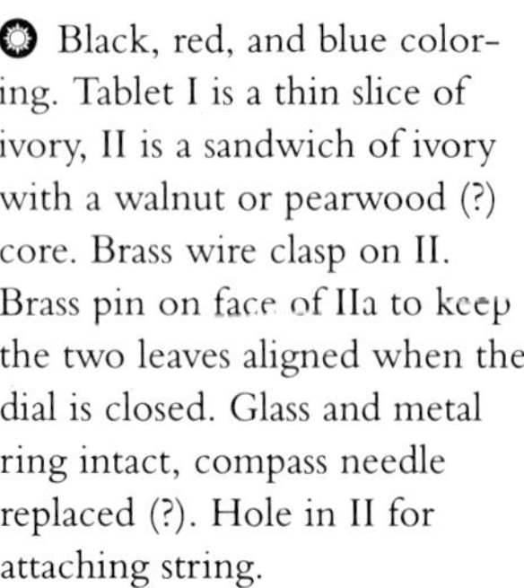

lb Vertical dial labeled VI–XII–VI in black numerals. Two small stamps.

lla Horizontal dial with a single scale for approximate latitude 47.5°, labeled 4–12–8 in black numerals. Cardinal directions given in compass bowl as SE, OR, ME, and OC. Magnetic declination 20° west of north. Two small decorative stamps.

llb Undecorated.

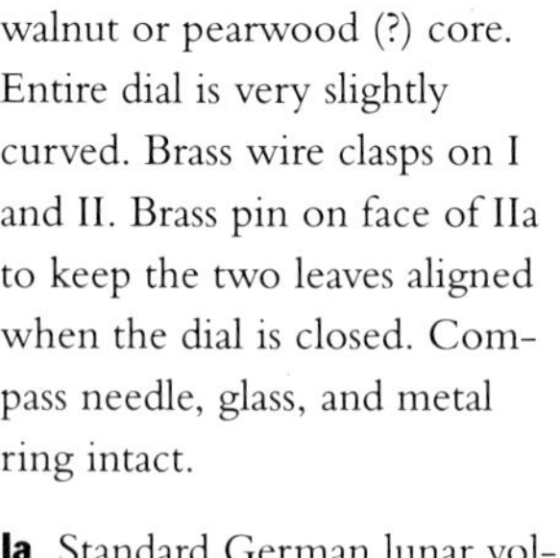

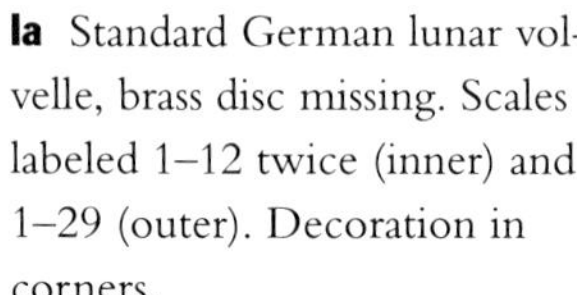

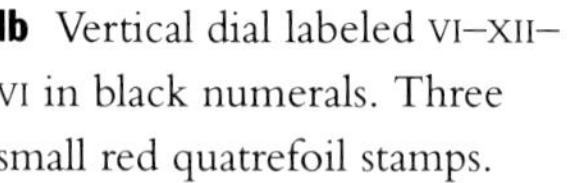

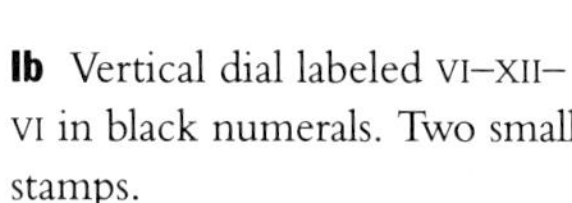

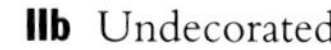

⚙ Black and red coloring. Tablet I is a thin slice of ivory, II is a sandwich of ivory with a walnut or pearwood (?) core. Entire dial is very slightly curved. Brass wire clasps on I and II. Brass pin on face of IIa to keep the two leaves aligned when the dial is closed. Com-pass needle, glass, and metal ring intact.

la Standard German lunar vol-velle, brass disc missing. Scales labeled 1–12 twice (inner) and 1–29 (outer). Decoration in corners.

lb Vertical dial labeled VI–XII–VI in black numerals. Three small red quatrefoil stamps.

lla Horizontal dial with a single scale for approximate latitude 48°, labeled 4–12–8 in black numerals. Maker's mark ("B") in compass bowl. Cardi-nal directions given in compass bowl as SEPT, ORI, MERI, and OCCI. Magnetic declination 20° west of north. Four small decorative stamps.

llb Undecorated.

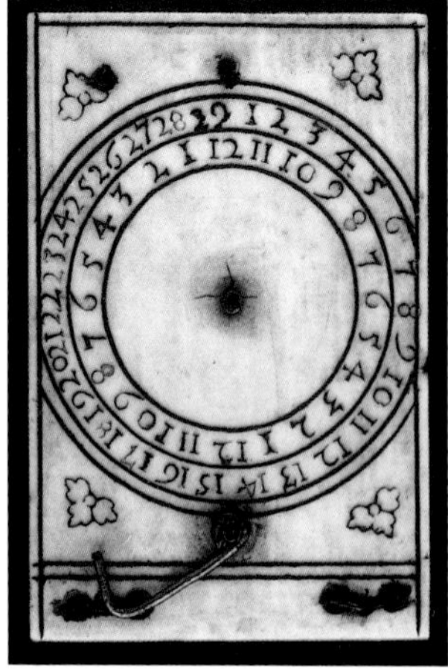

la
llb

42 Miniature Ivory and Wood Diptych

signed with maker's mark
[Karner workshop]
first half of 18th century
Nuremberg, Germany
2.8 (w) x 4.4 (l) x 1.0 (h) cm
Formerly David P. Wheatland
Collection (purchased in
France, 1952)
Inventory No. 7547

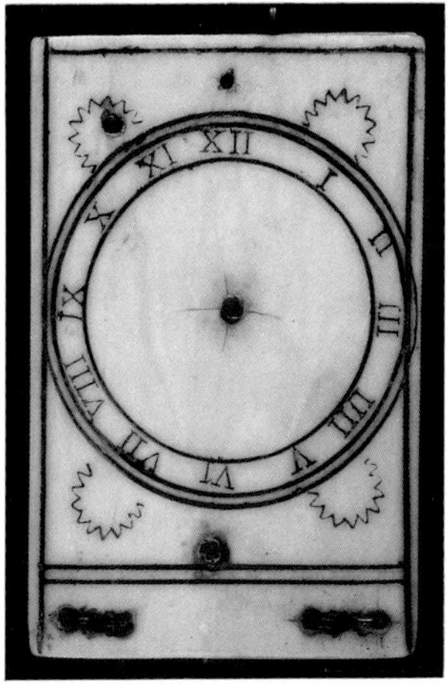

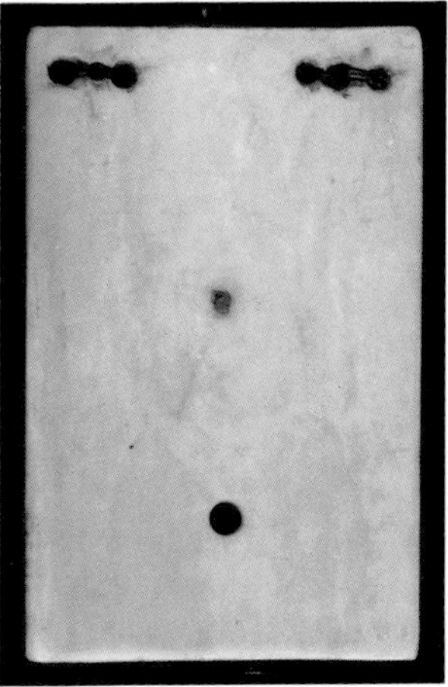

⚙ Black, red, and blue coloring. Tablet I is a thin slice of ivory (slightly curved), II is a sandwich of two slices of ivory with a pearwood (?) core. Brass wire clasp on II. Brass pin on face of IIa to keep the two leaves aligned when the dial is closed. Compass needle, glass, and metal ring missing. Hole in II for attaching string.

Ia Volvelle with only one scale from I to XII once, brass disc missing. Decoration in corners.

Ib Vertical dial labeled VI–XII–VI in black numerals. Three small red quatrefoil stamps.

IIa Horizontal dial with a single scale for approximate latitude 48.5°, labeled 4–12–8 in black numerals. Maker's mark ("B") in compass bowl. Cardinal directions given in compass bowl as SE, OR, ME, and OC. Magnetic declination 20° west of north. Two small red decorative stamps.

IIb Undecorated.

43 Miniature Ivory and Wood Diptych

signed with maker's mark
[Leonhard Andreas (?) Karner]
c. 1730
Nuremberg, Germany
3.3 (w) x 4.8 (l) x 1.5 (h) cm
Formerly Drecker Collection
No. 138, then David P.
Wheatland Collection
Inventory No. 7545

44 Miniature Ivory and Wood Diptych

signed with maker's mark
[Leonhard Andreas (?) Karner]
c. 1730
Nuremberg, Germany
3.4 (w) x 4.6 (l) x 1.4 (h) cm
Formerly Ernst Collection
No. 87
Inventory No. 7887

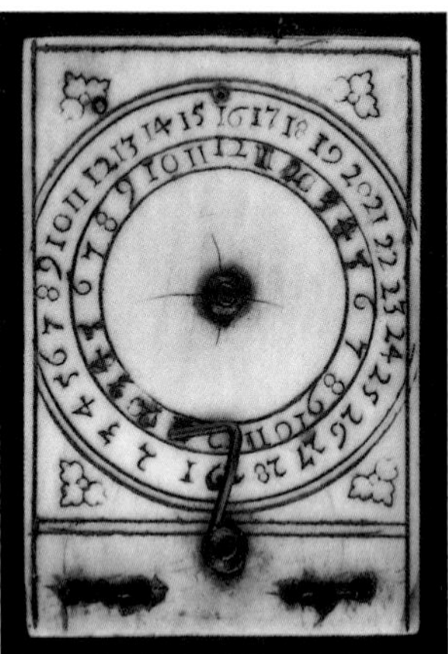

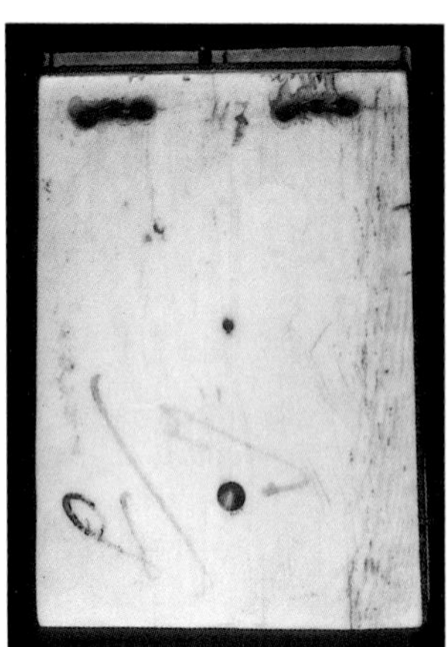

Ia | Ib
IIb | IIa

☀ Black and red coloring. Tablet I is a thin slice of ivory (outer surface slightly curved), II is a sandwich of two slices of ivory with a pearwood (?) core. Brass wire clasps on I and II. Brass pin on face of IIa to keep the two leaves aligned when the dial is closed. Compass needle, glass, and metal ring intact.

Ia Standard German lunar volvelle, rotating disc missing. Scales labeled 1–12 twice (inner) and 1–29 (outer). Decoration in corners.

Ib Vertical dial labeled VI–XII–VI in black numerals. Five small red decorative stamps.

IIa Horizontal dial with a single scale for approximate latitude 48°, labeled 4–12–8 in black numerals. Maker's mark (a hand) in compass bowl. Cardinal directions given in compass bowl as SEPT, ORIE, MERI, and OCCI. Magnetic declination about 20° west of north. Three small red decorative stamps.

IIb Undecorated. Numbered 47 in red ink near hinge. Several faint letters in black ink.

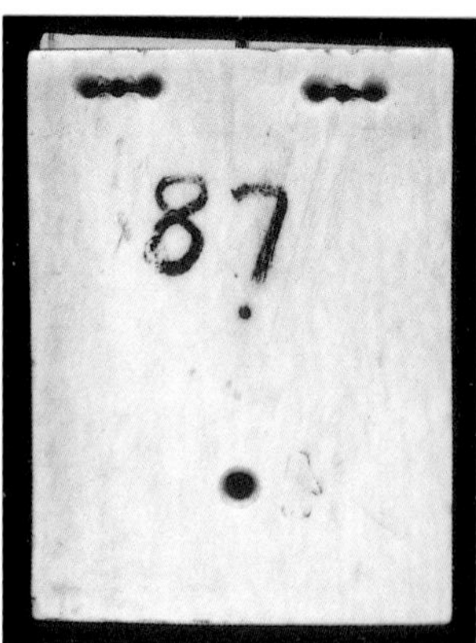

Ia
IIb

☀ Black, red, and blue coloring. Tablet I is a thin slice of ivory, II is a sandwich of ivory with a pearwood core. Brass wire clasps on I and II. Brass pin on face of IIa to keep the two leaves aligned when the dial is closed. Compass needle, glass, and metal ring intact.

Ia Standard German lunar volvelle with a stamped brass rotating disc (2.0 cm) with man-in-the-moon motif and index. Scales labeled 1–12 twice (inner and middle) and 1–29 (outer). Decoration in corners.

45 Miniature Ivory and Wood Diptych

signed with maker's mark
[Leonhard Andreas (?) Karner]
with original case (not shown)
c. 1730
Nuremberg, Germany
3.1 (w) x 4.5 (l) x 1.2 (h) cm
Given to David P. Wheatland
by Martha K. Wheatland, 1943
Inventory No. 7548

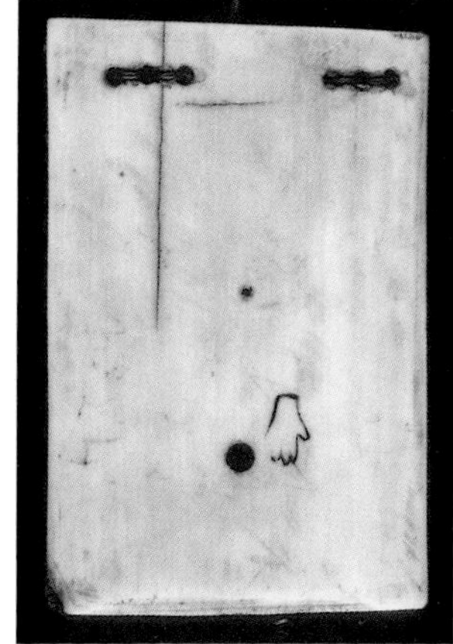

◉ Black, red, and blue coloring. Tablet I is a thin slice of ivory, II is a sandwich of two slices of ivory with a beechwood core. Front and back faces of ivory are noticeably curved. Brass wire clasps on I and II. Brass pin on face of IIa to keep the two leaves aligned when the dial is closed. Compass needle, glass, and metal ring intact. Includes black tooled leather case [3.7 (w) x 5.5 (l) x 2.1 (h) cm] which slides open and shut. Inner lining of multi-colored paper.

la Standard German lunar volvelle with a stamped brass rotating disc (1.7 cm) with index. Scales labeled 1–12 twice (inner and middle) and 1–29 (outer). Man-in-the-moon motif on volvelle. Decoration in corners.

lb Vertical dial labeled VI–XII–VI in black numerals. Two small red decorative stamps.

lla Horizontal dial with a single scale for approximate latitude 48°, labeled 4–12–8 in black numerals. Cardinal directions given in compass bowl as SEPT, ORIE, MERI, and OCCI. Magnetic declination about 20° west of north. Three small red decorative stamps.

llb Undecorated except for maker's mark, a hand.

lb Vertical dial labeled VI–XII–VI in black numerals. Two small red decorative stamps.

lla Horizontal dial with a single scale for approximate latitude 49°, labeled 4–12–8 in black numerals. Cardinal directions given in compass bowl. Magnetic declination 20° west of north. Two small red decorative stamps.

llb Plain except for maker's mark, a hand, near the bottom. Numbered 87 in black ink.

46 Rectangular Ivory Diptych

signed with maker's mark [Michael Lesel]
early 17th century
Nuremberg, Germany
7.1 (w) x 9.1 (l) x 1.3 (h) cm
Formerly Drecker Collection No. 112,
then David P. Wheatland Collection
Inventory No. 7559

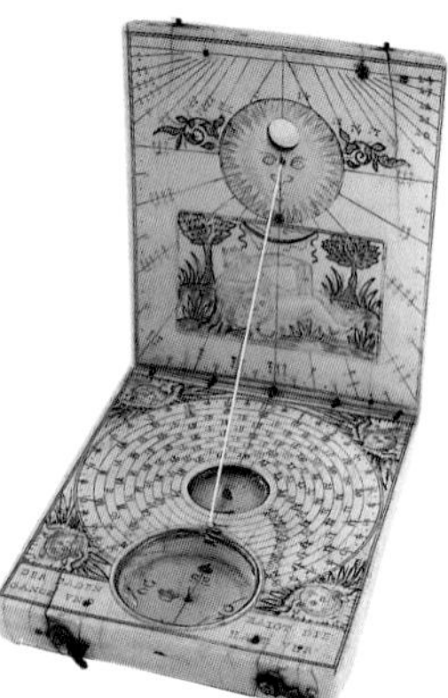

○ Black, red, brown, and green coloring.
Compass viewing-hole in I with intact
mica held in place by a metal ring, slot for
wind vane in side of II. Brass clasps on I
and II, four bun feet on II. Braided silver
ring in central hole, copper ring and glass
over compass bowl.

Ia Wind rose with 32 points, numbered
1–32 beginning at east, 16 directions
labeled or abbreviated in German, hand-
shaped brass index, wind vane missing.
Crack in I held together with rectangular
brass patch, lesser crack in II. Four-winds
decoration in corners.

Ib Two pin gnomon dials (gnomons missing) for Italian hours (right, labeled 14–24
in black numerals) and Babylonian hours
(left, labeled I–X in red numerals). Large
scowling sun-face with foliate decoration.
Vertical dial labeled VI–XII–VI in faint black
numerals, with lines and dots marking the
half- and quarter-hours. Engraving of a
nude person with an hour glass and a skull
with a snake slithering through the eye
socket.

IIa Horizontal dial for approximate latitude 48°, with 10 scales (the two outermost
being identical), ranging from 16–24 (innermost) to 8–24 (next to outermost) and
4–12–8 (outermost), alternately in black
and red numerals in successive scales, with
lines and dots marking the half- and
quarter-hours, respectively, in the outer
scale. The outer scale indicates equal common hours, and the inner scales Italian
hours. (See page 18 for an explanation of
these scales.) Compass bowl with crown
maker's mark. Cardinal directions given in
compass bowl as SE, OR, ME, and OC. Magnetic declination 4° east of north. "DER
FADEN ZAIGT DIE GANCZ VND HA[L]B VHR"

[the thread shows the whole and half-
hours]. Four-winds motif. Mica-covered
hole at center of dial, 1.85 cm in diameter.

IIb Standard German lunar volvelle with a
stamped brass rotating disc (2.4 cm) with
index. Scales labeled 1–12 twice (inner and
middle) and 1–29 (outer). Man-in-the-

moon motif on brass volvelle disc. Scale
outside lunar volvelle indicates the
correction in hours and minutes between
apparent lunar and solar time. Crown
maker's mark appears once near hinge.
Four-winds motif. Numbered 27 in red
near hinge. Image inverted for legibility.

3

French Diptych Sundials

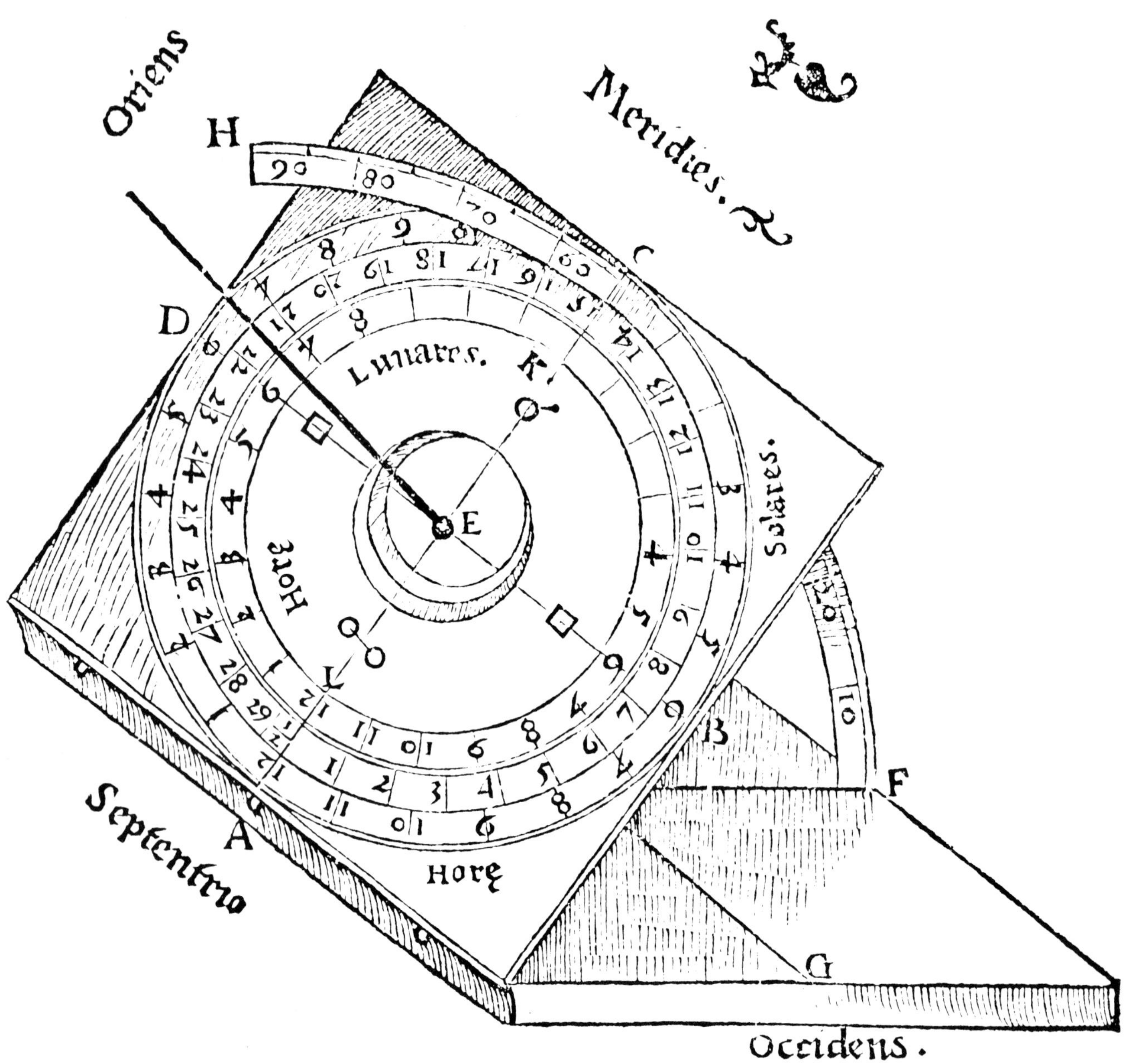

Oriens
Meridies.
Septentrio
Occidens.
H
D
A
C
B
F
G
E
Lunares.
Horæ
Horæ
Solares.
90 80 70
9
8
7
6
5
4
3
2
1

Historical Introduction

by A. J. Turner

From the variety of forms and styles employed by the makers of ivory diptych[1] dials in France the following typology can be constructed:

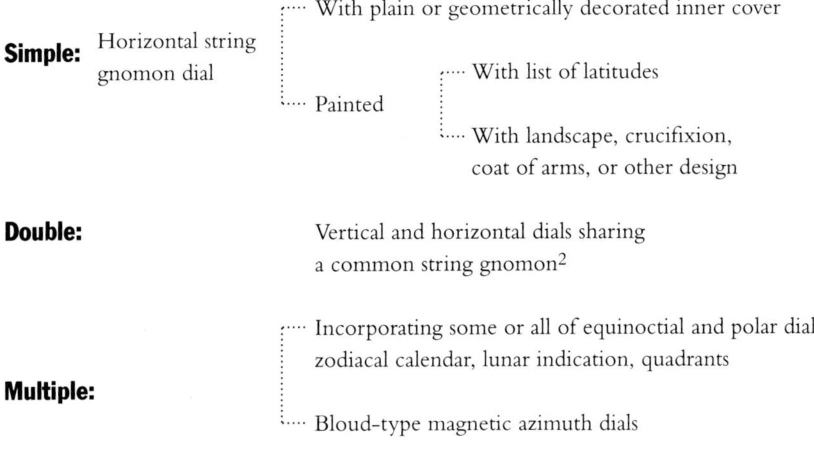

Simple: Horizontal string gnomon dial
- With plain or geometrically decorated inner cover
- Painted
  - With list of latitudes
  - With landscape, crucifixion, coat of arms, or other design

Double: Vertical and horizontal dials sharing a common string gnomon[2]

Multiple:
- Incorporating some or all of equinoctial and polar dials, zodiacal calendar, lunar indication, quadrants
- Bloud-type magnetic azimuth dials

That this typology corresponds in whole or in part with the development of ivory diptych sundials in France we do not know, for French dials have never been studied in detail and very little information concerning them is available. The remarks that follow are therefore merely notes arising from a rapid preliminary survey. If they seem to lead to any conclusions, these must be considered as tentative.

The primary problem in the study of French ivory diptych dials is that, unlike their Nuremberg counterparts, they are rarely dated and still more rarely signed. Stylistically they seem to fall into three main groups. One is composed of simple string gnomon dials with only a small amount of decoration, which is usually geometrical patterns inscribed by dividers or a graver and straight edge. A second group is characterized by painted decoration. The third is distinguished by a far greater homogeneity of decoration and layout of the dials than is seen in the first two groups. In particular, the dials in this group regularly use a characteristic border of repeating stylized trilobe elements (see cat. nos. 72–75).[3] A higher percentage of dials in this third group are signed than in the other two, always by Dieppe makers. It is therefore reasonable to ascribe unsigned dials of similar style to the same Dieppe "school." Dials in the other two groups are more difficult to place although, since the rare signed examples are signed by Paris makers and there are strong stylistic similarities with several

1. *Diptych* is here used in the loose general sense of a dial composed of two leaves or tablets hinged to each other so that they open at a right angle. Properly defined, a diptych dial carries on its two inner surfaces respectively a vertical and a horizontal dial which share a common string gnomon. For some discussion, see Gouk (1988), p. 10; Bryden (1988), introduction to section I, 4.2; and Section 1, above.

2. It is this group and some of the multiple dials which alone fulfill the conditions of the strict definition of diptych dials. Among French dials there are astonishingly few examples; see Bryden (1988), no. 72.

3. Named, not inappropriately, by Bryden (1988) the "pecked border pattern."

Diptych sundial with a lunar volvelle from Oronce Fine, *Protomathesis*, Paris, 1532, f. 178. By permission of the Houghton Library, Harvard University.

of the unsigned dials, it seems reasonable to postulate a small but distinct production of ivory dials in Paris.

Ivory-working in Paris has a long history. Already in the *Livre des métiers* of 1293[4] four groups of specialist craftsmen working in ivory could be distinguished. These were the "patenôtriers," who made ivory buttons and ivory birdcages as a sideline; the "couteliers faiseurs de manches," who made handles for swords, hallebards, scissors, surgical and mathematical instruments, knives, and penknives, many of which were in ivory; the "deiciers," who made dice, chessmen, and pieces for other games in ivory but who were absorbed into the patenôtriers early in the sixteenth century;[5] and the "tabletiers," who in the thirteenth century made writing tablets but by the sixteenth century worked more generally in ivory and combined with the other groups to form the "peigniers-tabletiers-tourneurs et tailleurs d'ymages en yvoire."[6] Each group, however, continued its own speciality.

The most likely group to have made ivory diptych dials is the tabletiers, whose basic product, writing tablets, closely resembles diptych dials in form in being hinged together, and no doubt decorated, in a similar way. Perhaps the makers called upon the "ymagiers-peintres," who specialized in ivory-painting, to decorate those dials which carried elaborate painted scenes. For one important workshop, that of Julian Thomas, in which sundials were made along with games and counters, an inventory exists carried out in 1598.[7] But examination of the making of ivory dials in Paris is hindered by the fact that the names of only three makers are known from existing dials. These are Pierre Dujardin, Dujardin le jeune, and A. André.

Pierre Dujardin is known by four instruments dated 1614, 1617, 1620, and 1627.[8] Dujardin le jeune, whose relationship with Pierre Dujardin is not established, is known by only one instrument dated 1648, and A. André by two.[9] Between the work of the three makers there are close stylistic similarities, as there are between their signed pieces and several unsigned instruments. To make any attributions on the basis of such a small sample would be unwise, but the similarities between the ivory dials by Dujardin and André and several of the unsigned examples are so close that there can be no doubt of their origin in the same place and probably in a very small number of workshops which themselves perhaps drew upon a common source.[10] Most of the dials that can be ascribed to this group, whether simple, double, or multiple, share use (if they are rectangular) of a plain frame made up of two border lines connected at the corners by an internal diagonal; they also share vertical latitude scales on face Ib, which are either set along one edge of the instrument and work with a strut housed in IIa, against which the lid can be inclined when an equinoctial dial is marked, or placed centrally with a spring adjustment for the string gnomon.[11] The form of the punched and engraved letters across all the dials is closely similar, as is the characteristic rounded and rather uneven script written in gold on a blue, green, or red background employed for latitude or other lists and inscriptions on several of the dials. A similar script is used on dials of different shapes, square, rectangular, oval, or hexagonal.

4. L'Espinasse and Bonnardot (1879).

5. Franklin (1906), pp. 257–258.

6. 30 July 1507; an arrêt du Parlement of 23 January 1579 following a dispute within the combined company pronounced in favor of the combined corporation. L'Espinasse (1886–1897), vol. 2, p. 670. For the statutes of the corporation, see vol. 2, pp. 676–679.

7. Archives Nationales, Paris, Minutier centrale, Et. XXXIV, 24.

8. These are listed with references in Chapiro, Meslin-Perrier, and Turner (1989), p. 96.

9. The Dujardin le jeune instrument is in the Boerhaave Museum, Leiden Inv. 9945. Tardy (1972) describes him as "horloger-tabletier" in 1645 but cites no evidence. If, as use of the phrase "le jeune" suggests, his Christian name was also Pierre, then he may be the Pierre Dujardin "maitre peigner tablettier" who on 5 December 1650 married Pierrette Grillot. In the marriage contract (Archives Nationales, Paris, Minutier centrale, Et. XXXV, 62) he is described as being twenty-five years old, a native of Claye near Meaux, and the son of Claude Dujardin, vigneron. In this case the elder Pierre Dujardin should be an uncle or cousin. For the instruments by André, one is in the Harvard Collection (see cat. no. 55), and the other is a portable horizontal dial sold in 1935, of which the present whereabouts is unknown. See Schuhmann (14 & 15 novembre 1935), lot 123. An ebony diptych dial now in the Museum of the History of Science, Oxford (F. 121), which is marked with a crowned 𝔄, may also be attributable to A. André. See also cat. no. 49.

10. The unsigned multiple dial dated 1614 in the Whipple Museum, Cambridge (see Bryden, 1988, no. 72), has on face Ia an equinoctial and polar dial diagram that is identical with that on face Ia of the multiple dial made by Pierre Dujardin in 1627 in the Musée de la Renaissance, Ecouen (Chapiro, Meslin-Perrier, and Turner, 1989, no. 77, p. 97). Face IIb is similarly close in style, while the geometric decoration recalls that of a simple string gnomon dial illustrated in Wynter (1974), p. 4.

11. This spring adjustment arrangement and the double frame are shown clearly on the dial dated 1623, cat. no. 51.

The oval dials tend to carry geometrical decoration or to be painted. Dated examples of oval dials come from the late sixteenth century,[12] and although simpler they are reminiscent, particularly in their use of double incised lines and half-moon decorations, of some of the dials of Hans Ducher II or III. That there was some direct influence is possible, but one might equally postulate that the similarity between Ducher dials and the French dials, if it is not merely coincidence, comes from their both having developed from a common source. In this context it would be interesting if an ivory dial excavated in the Louvre could be accurately dated.[13] Although ascribed to a Nuremberg maker by its excavators, the plainness of the dial, the simplicity of its style and layout, and the form of its script suggest rather that it is a French example, perhaps an early one.

In 1612 Justus Zinzerling noted of Dieppe that "they make here many things out of bone, whalebone and ebony."[14] Three decades later John Evelyn remarked that Dieppe "exceedingly abounds in workemen that make and sell curiosities of Ivory and Tortoise-shells, in which they turn, and make many rare toyes."[15] The industries that Zinzerling and Evelyn noted were relatively newly established, for although ivory had been imported from West Africa to Dieppe since at least the fourteenth century, in general it was passed straight on to Rouen and Paris. That in 1518 two dozen box-shaped dials worth forty-two sols had made up part of a shipment from Dieppe to England may imply local manufacture but need not.[16] Only toward the end of the sixteenth century did the Dieppe ivory manufacture begin to develop, although it had clearly done so to good effect by the time of Evelyn's visit, when craftsmen in all branches of ivory-working were apparently numerous. By the end of the seventeenth century they were predominant in the kingdom. As Savary de Bruslons noted, in the early eighteenth century Dieppe ivory-workers were reputed to be the best in France.[17] They first imitated and then surpassed the work of their Paris counterparts, completely taking over the trade. In the process they also took over the making of sundials. That by 1662 attempts were being made in Dieppe to protect the price of dials made in brass may suggest that the late 1650s and early 1660s was the period when ivory dial-making first became prominent in Dieppe. One of the craftsmen who signed the price-fixing agreement, Nicolas Crucefix, also made ivory dials, one of which has survived. At least one other member of the group, Jacques Lemoyne, also worked in ivory.[18] In 1668 Charles Bloud, Gabriel Bloud, Ephraim Senecal, Jacques Senecal, and Francois Saillot are noted as "quadranniers," and an act of tabellionage of 1673 adds two further names, David Postel and Mathieu Berville.[19] Other names, like that of Jacques Guerard,[20] are perhaps still to be added from the dials themselves or from other sources, but it is unlikely that the total number of specialist dial-makers was ever very large.

The ivory dials of Dieppe, as noted above, are more frequently signed than are their Paris-made counterparts of the first half of the century, and they share a more homogeneous style. Characteristic are the "pecked" and leaf-design borders and a symmetrical but quite free-flowing foliate decoration. Neither design is engraved or punched. Both are applied with tools similar to those used by book-binders onto the heat-softened surface of

12. For example, cat. no. 47, which is dated 1584.

13. Illustrated in Trombetta (1987), p. 139.

14. *Itinerarium Galliae*, 1612, "varia hic conficiuntur ex ossibus, balenarum et ebone."

15. De Beer (1955), vol. 2, p. 124.

16. Anthiaume (1920), vol. 2, p. 224.

17. *Dictionnaire de commerce*, 1723.

18. Le Corbeiller (1914), p. 64.

19. Milet (1904), p. 22; (1906), p. 6.

20. Signature on a magnetic azimuth dial, Whipple Museum, Cambridge (Bryden, 1988, no. 118). Presumably Jacques Guerard belonged to the family of the cartographer and teacher of navigation Jean Guerard, who worked a generation earlier. See Milet (1904), p. 19. See also cat. no. 62.

the ivory.[21] But although distinctive, the layout of Dieppe dials and some residual decorative elements show quite clearly that there is, at least stylistically, continuity between the Paris-made dials of the early seventeenth century and the Dieppe instruments of the second half. The double- and single-lined frames with internal corner diagonal mentioned above occurs on dials from both areas,[22] sometimes carrying the "pecked" border inside it on Dieppe-made examples.[23] If an equinoctial dial is incorporated, the latitude scale is almost invariably placed along the right-hand edge, as in the earlier Paris dial having a brass strut to support the lid.[24] Several of the Dieppe dials incorporate a volvelle like the Paris-made dials in the center of face Ib. Others, non-equinoctial, use the central latitude scale with a spring-adjustable attachment for the string gnomon, simple pendant plumb bobs which can also be found on Paris dials, and a script which although engraved, not written, is a round cursive clearly related to that used on the Paris group.[25] Particularly eloquent as a witness of this continuity, even transmission, from Paris to Dieppe is a large compendium by Charles Bloud à Dieppe which not only incorporates a nocturnal like the compendia of Dujardin but has a gilt volvelle set into face Ib with a blue painted surround.[26]

If the stylistic evidence of the dials themselves thus strongly suggests continuity, even transmission, from Paris to Dieppe in the making of ivory dials, there is as yet no documentary evidence to confirm it. Of the origins of Charles Bloud and his brother Gabriel, or of Ephraim Senecal and his, nothing is known. It is just as likely that they were Paris-trained workmen who migrated to Dieppe in the mid-century as that they were born in Dieppe. Whatever the case, the manufacture of dials in Dieppe was formed and given identity by their work. With the development, probably by Charles Bloud, in the mid-century of an ingenious but simple-to-use multiple dial, a completely individual, and, as it proved, very popular dial, was placed on the market.

The magnetic azimuth dial has become so closely identified with Dieppe manufacture as to seem characteristic, the other Dieppe dials being merely simpler versions, without a magnetic azimuth dial, using the same style. In effect, if the foregoing account be accurate, we can see that exactly the opposite is true. The form of the French diptych dial was already well established by the time that it became possible to use magnetic meridians for marking the time, and the new element was simply incorporated into the older form. The development of magnetic azimuth dials depended upon the fact that toward the middle of the seventeenth century, magnetic declination was approaching zero (true zero at London, 1657–1661; true zero at Paris, 1662–1665).[27] This meant that during the middle decades of the century the magnetic meridians were close to or coincided with the geographical meridians. It was thus possible to construct an azimuth dial that did not require a strongly defined shadow to be cast in order to show the time. It could therefore be used in overcast conditions as long as one could determine the position of the sun and/or the moon.

Although it is clear that magnetic azimuth dials predate Charles Bloud and the Dieppe manufacture, Bloud's clear claim to have invented them, made in the "Invenit et fecit" formula that he used on several magnetic

21. Use of this technique on Dieppe dials was first noted by Michel (1939), p. 37. Cf. Murdoch (1984), p. 66.

22. See, for example, cat. nos. 57 and 58, both signed by Charles Bloud.

23. Whipple Museum, Cambridge, inv. 1676; Bryden (1988), no. 115.

24. A rare exception to this is found in a tortoise-shell magnetic azimuth dial by Charles Bloud now in the British Museum. See Ward (1981), no. 113, p. 45, where the scale is on the left-hand side. Another tortoise-shell dial is preserved in the Museum of the History of Science, Oxford.

25. An octagonal string gnomon dial by "C. Bloud A Dieppe" formerly in The Time Museum, Rockford, Illinois (see Christie's, 14 April 1988, lot 65), clearly illustrates these three elements.

26. Museum of the History of Science, Oxford F. 147.

27. Milet (1904), pp. 25–26, and see note 30, below.

azimuth dials but not on other forms of dials that he signed, does suggest that he may have been responsible for the precise form that such dials took. In "Bloud-type" dials the magnetic azimuth dial is combined with a specific form of perpetual calendar marked on the disc contained within the zodiacal calendar and with a luni-solar conversion volvelle so that the dial could be used at night. These elements were incorporated into the classic form of French diptych dial with equinoctial and polar dials and a horizontal string gnomon dial around the compass in the lower tablet. In this context, the fine dial made by Bloud for Henri de Longueville[28] may be of particular significance. That it is apparently the only Bloud dial known to be dated, and that its layout does not conform exactly to the classic pattern of magnetic azimuth dials, suggests that it may stand at the head of the series, the first fruit of Bloud's imagination produced for an influential patron.[29]

Exactly how long magnetic azimuth dials continued to be manufactured we do not know. Between 1662, when the magnetic declination at Paris was 0°, and 1684/5, when it was just over 4° west,[30] the increase was perhaps insufficient to make a noticeable difference to the reading obtained from a small sundial, particularly at a period when accurate long-pendulum clocks were only just coming into general use.[31] Thereafter, however, the continuing increase in the declination value (8°50' in 1702) must have led to a noticeable discrepancy even between the time shown by the magnetic azimuth dial and that given by the equinoctial or polar dials. From about 1680 or 1690 onwards one would expect the number of magnetic azimuth dials manufactured to have declined. That they did decline (along with the the making of ivory dials in general) was, however, the result of external causes. Bloud, Senecal, and many of the other Dieppe dial-makers were Protestant and as such were severely persecuted after the revocation of the Edict of Nantes in 1685.[32] By 1699 no one of the name of Bloud lived in Dieppe, and of the five children of Ephraim Senecal three had already abandoned the town by 1686. For a small trade the consequences could not but have been disastrous. The generation of Charles and Gabriel Bloud seems to have had no successors. By the early decades of the eighteenth century the making of ivory diptych sundials in France had largely disappeared.

28. Cat. no. 56.

29. Alternatively, the date may have significance only in the life of the patron. For a detailed discussion of the origins and nature of the magnetic azimuth dial, see Maddison and Turner (in preparation).

30. Le Monnier (1771), pp. 459–460. The corresponding values for London are 1657 0°, 1680 4° 8'; see Malin and Bullard (1981), pp. 370–371.

31. Although pendulum clocks had been available since 1658, the anchor escapement which dramatically improved their performance was not applied until about 1668/69 at the earliest.

32. Milet (1904), pp. 25–26. In the "Roole general Des nouveaux Convertis de la Ville de Dieppe leurs caracteres et Dispositions Desprit [*sic*]…", 1686, Bibliothèque Nationale, Ms mélanges Colbert 6, fols. 189ff, Gabriel and Charles Bloud (fols. 192r, 193r) are both described as "Converty mal," as are Jacques Seneschal (fol. 198v) and Effrain (Ephraim) Seneschal (fol. 235r) [another spelling for the family here named Senecal].

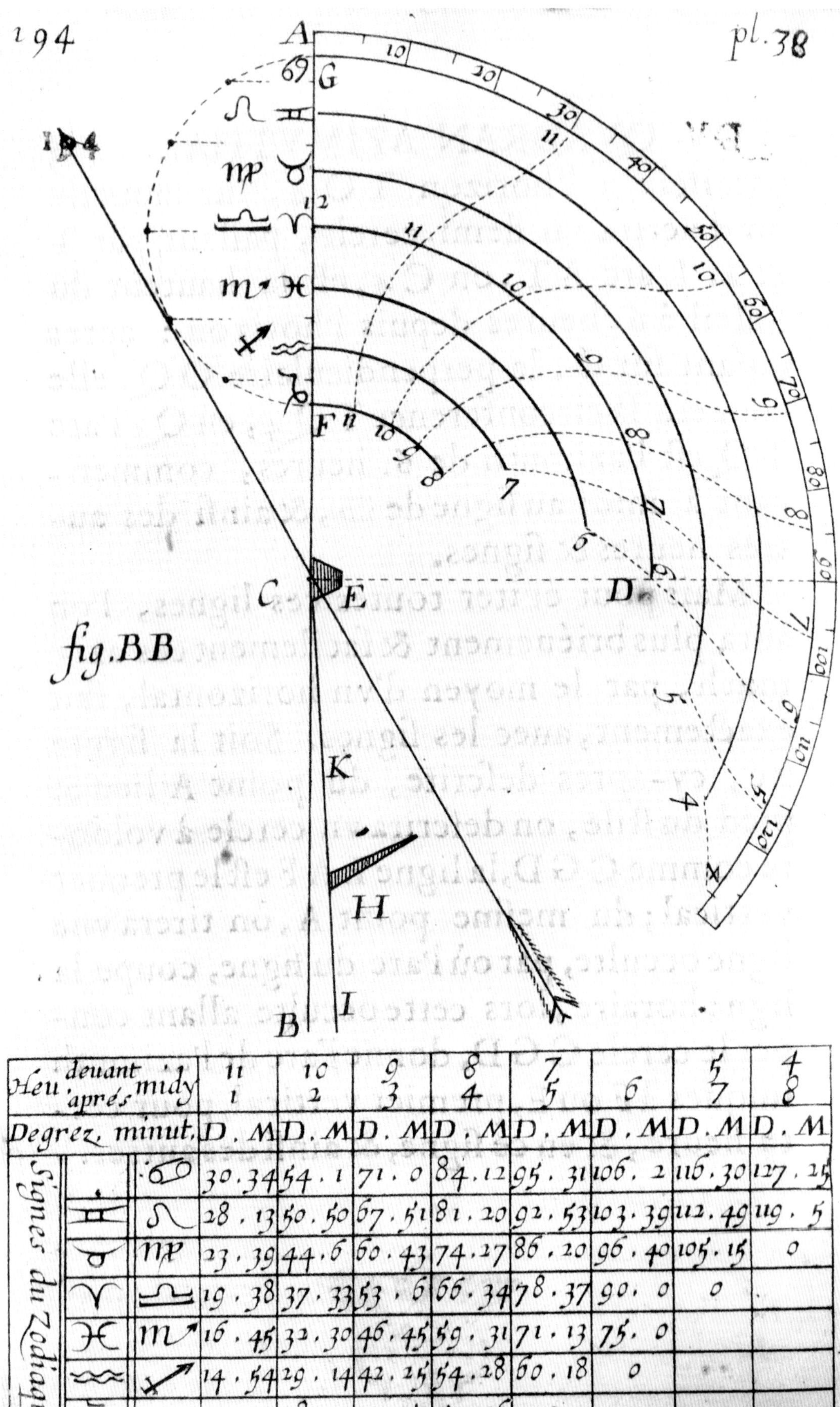

Heu. deuant / après midy	11 / 1	10 / 2	9 / 3	8 / 4	7 / 5	6	5 / 7	4 / 8
Degrez. minut.	D. M	D. M	D. M	D. M	D. M	D. M	D. M	D. M
♋	30.34	54.1	71.0	84.12	95.31	106.2	116.30	127.25
♊ ♌	28.13	50.50	67.51	81.20	92.53	103.39	112.49	119.5
♉ ♍	23.39	44.6	60.43	74.27	86.20	96.40	105.15	0
♈ ♎	19.38	37.33	53.6	66.34	78.37	90.0	0	
♓ ♏	16.45	32.30	46.45	59.31	71.13	75.0		
♒ ♐	14.54	29.14	42.25	54.28	60.18	0		
♑	14.20	28.4	40.50	52.36	0			

Planche 38.

Early French Diptych Sundials

The earliest French diptych sundial in the Harvard Collection, dated 1584, has two fixed string gnomon positions at 39 and 51 degrees of latitude, corresponding to Lisbon and London, respectively. A stylistic progression is noticeable in the Harvard Collection's dials dated 1584, 1603, 1623, 1624, and 1642. Common stylistic elements and gradual trends in design among cat. nos. 47, 48, 51, 52, and 53 suggest that they may have originated in the same workshop. The lack of mention of Dieppe in the latitude tables on these dials, as well as the documented presence of several ivory sundial makers in Paris before 1650, suggests that there may have been one or more workshops operating simultaneously in Paris throughout the early seventeenth century.

There is perhaps a stylistic connection between the earliest French ivory diptychs and the oval diptychs produced by Hans Ducher ca. 1590. These diptychs, made of ivory or composite ivory and wood, are characterized by stylized circular and oval motifs (similar to those used by the Duchers), the use of red Roman numerals on the lower dial plate, sometimes a lower tablet much thicker than the upper one, and occasionally a painted image on face Ib.

A distinguishing feature of the French diptychs is the use of brass or paper compass bowl inserts. Brass inserts are encountered only in the earliest French diptychs. The paper inserts were often labeled with the latitudes of various cities (most of them French); occasionally one finds small grid-like gazeteers which, in an abbreviated fashion, list the various attributes of the cities other than just their latitudes. For example, a list may include a city's status within the hierarchy of the Roman Catholic Church (such as whether it had a bishop, archbishop, or a cardinal to its credit). The paper inserts were often painted bright colors.

Another feature of early French diptychs is the inclusion of a sliding mechanism to adjust the angle of the string gnomon (and thus the latitude at which one may observe the time); in contrast, Nuremberg diptychs provide one or more holes at fixed latitudes through which one can attach the string gnomon. The sliding scales on the French dials are often calibrated for use between 40 and 50 degrees of latitude.

Early form of the magnetic azimuth sundial from Dom Pierre de St. Marie Magdeleine, *Traitte d'hologiographie*, Paris, 1645, page 194. Private Collection.

47 Oval Ivory Diptych

unsigned
dated 1584
France (?)
4.2 (w) x 5.6 (l) x 1.6 (h) cm
Formerly Drecker Collection No. 212,
then David P. Wheatland Collection
Inventory No. 7493

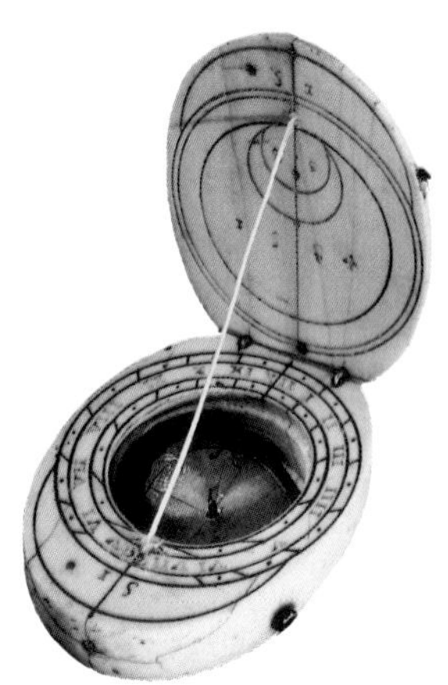

⚙ Red and black coloring. Mixture of
silver and copper hooks, eyes, and hinges
on I and II. The upper plate, considerably
thinner than the bottom, is broken lengthwise. The brass insert in the compass bowl
is twisted.

Ia Undecorated.

Ib Decorated with a motif of circles,
ovals, and a crescent. Holes provided for
the string gnomon labeled 39° and 51°.
Inscribed 1584.

IIa Horizontal dial with two hour scales
numbered V–XII–VII in red numerals and
labeled for 39° (inner circle) and 51° (outer
circle) latitude, with dots marking the half-
hours. Brass compass bowl insert labels the
four cardinal directions as M, P, T, and L,
with an arrow indicating a magnetic declination of 17° west of north. Oval overlaps
compass bowl.

IIb Undecorated. Marked 217 in black ink.

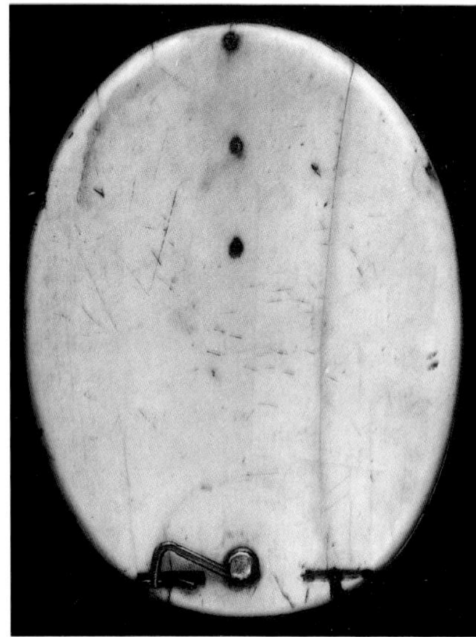

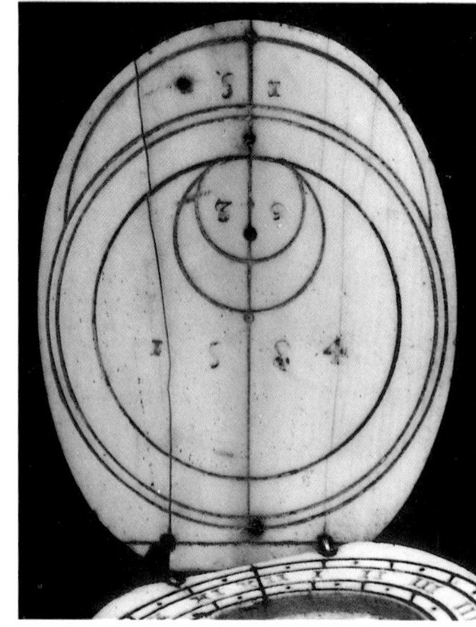

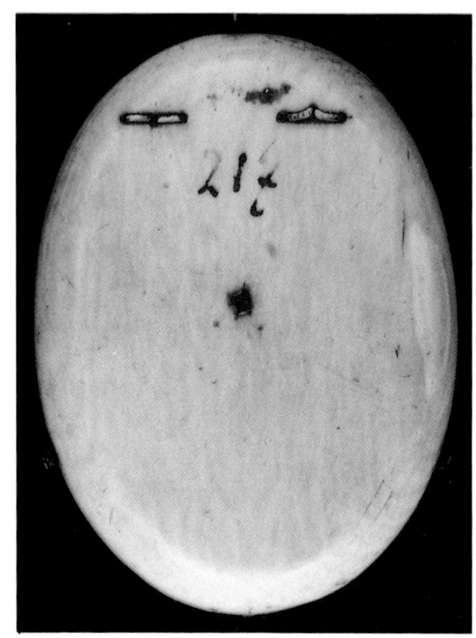

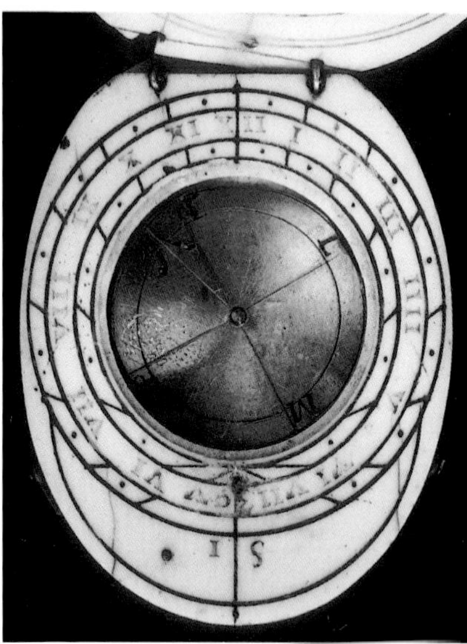

<table>
<tr><td>Ia</td><td>Ib</td></tr>
<tr><td>IIb</td><td>IIa</td></tr>
</table>

48 Oval Ivory Diptych

unsigned
late 16th century
France (?)
4.6 (w) x 6.0 (l) x 1.1 (h) cm
Formerly Harold Gillingham Collection
No. 104, then David P. Wheatland
Collection
Inventory No. 7522

⚙ Red and black coloring. Silver hook
clasp on I, copper eyes on I and II, brass
hinges. Brass insert in the compass bowl.
Both outer faces are somewhat discolored.
Tarnished brass ring to secure compass
glass. Glass is a replacement.

Ia Undecorated.

Ib Decorated with a painted floral motif,
including a red flower and stems with green
petals on a blue-gray background. Much of
the design has worn away.

IIa Horizontal dial with a single scale for
approximate latitude 47.5°, labeled V–XII–VII
in red numerals. Brass compass bowl insert
labels the four cardinal directions as SEPT,
ORIE, MERI, and OCCI [north, east, south,
and west]. Magnetic declination 13° east
of north. An ornate blued-steel compass
needle (later?) has a small jewel at its center.

IIb Undecorated. Marked "German
c. 1700/ 1925 Campbell 100" on modern
paper label.

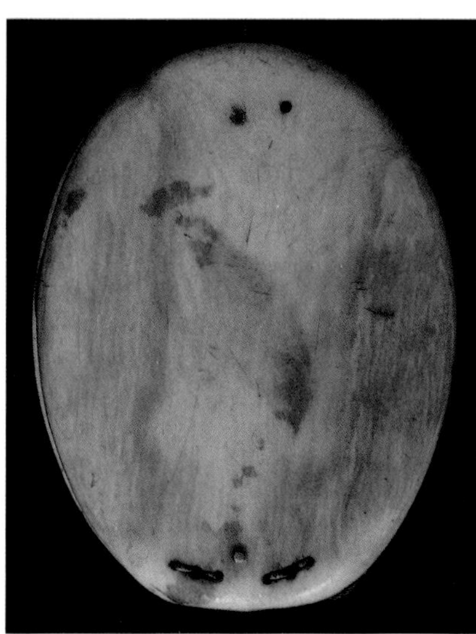

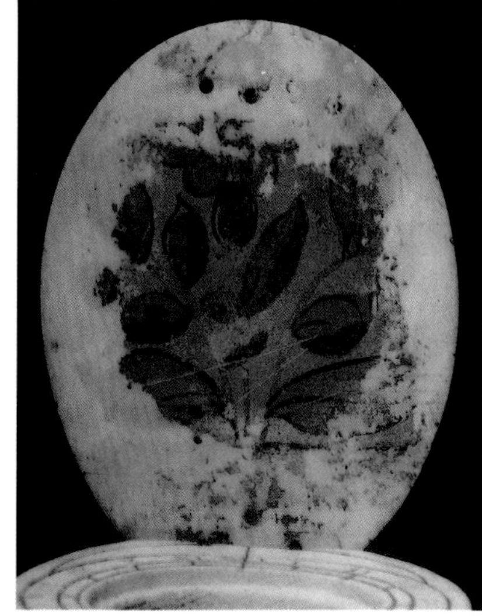

<table>
<tr><td>Ia</td><td>Ib</td></tr>
<tr><td>IIb</td><td>IIa</td></tr>
</table>

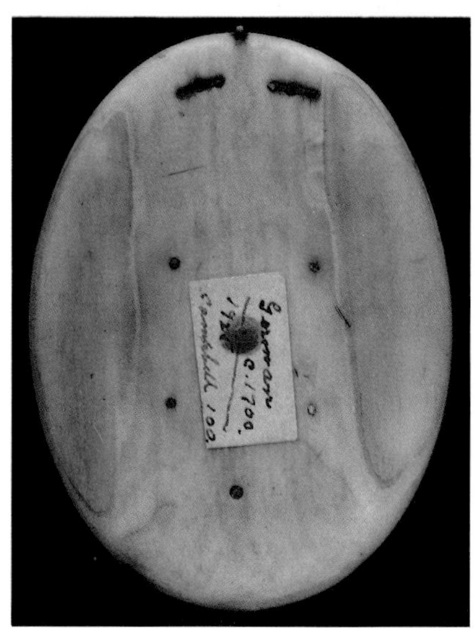

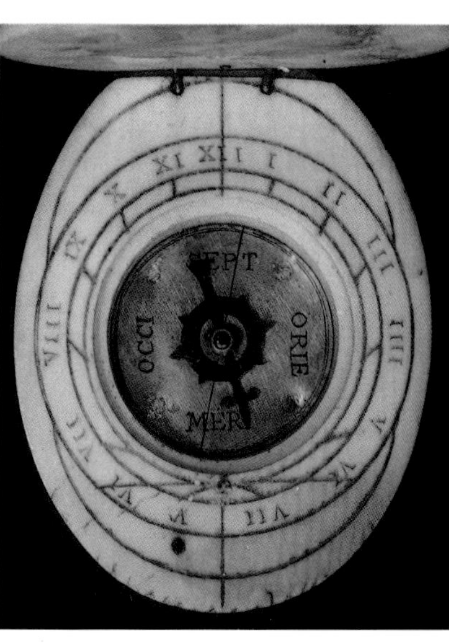

49 Oval Ivory and Wood Diptych

signed with maker's mark [A. André (?)]
dated 1603
Paris (?), France
6.1 (w) x 7.7 (l) x 1.5 (h) cm
Formerly Harold Gillingham Collection
No. 102, then David P. Wheatland
Collection
Inventory No. 7504

⊘ Red and black coloring. Silver hooks, eyes, and hinges on I and II. The upper tablet is a thin slice of rosewood, and the bottom is a composite of ivory and rosewood. The bottom of the compass bowl has a paper insert, with pointers painted red and blue. A thin, braided silver ring holds the glass over the compass bowl in place. Blued-steel compass needle with gilt brass center.

Ia Undecorated. Four holes for the string gnomon labeled 42°, 45°, 48°, and 50°.

Ib Latitude table of 12 cities, most of them French (42°–50° latitude), painted in gold on a red background. Dated 1603. Gold foliate ornamentation.

IIa Horizontal dial with three hour scales, outer scale labeled IIII–XII–VIII, middle scale unlabeled, and inner scale labeled VI–XII–VI in red numerals for 42°, 45°, and 48°, respectively, with dots marking the half-hours. Paper insert in the compass bowl rotated to an angle of 21° west of north (possibly a replacement). Typical French compass needle. Double-lined oval motif surrounds circular compass bowl.

IIb Undecorated. Two small screws (added later?). Maker's mark appears to be the letters V and A superimposed, or perhaps two A's with a small crown above (likely the mark of A. André of Paris). See page 100, note 9, above.

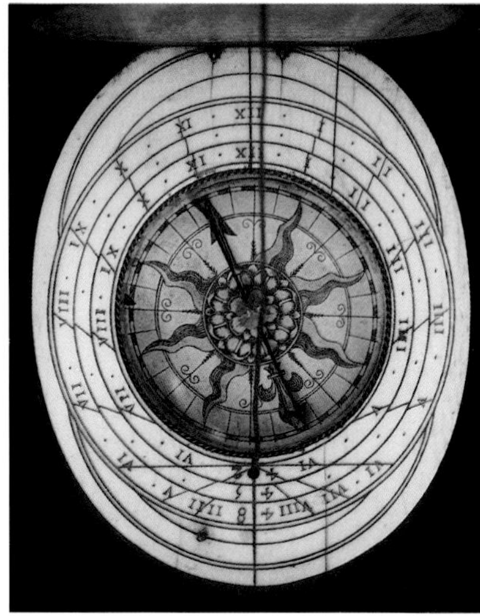

Ia | Ib
IIb | IIa

50 Octagonal Ivory Diptych

unsigned
dated 1624
France (?)
6.3 (w) x 6.5 (l) x 2.5 (h) cm
Formerly Ernst Collection No. 41
Inventory No. 7841

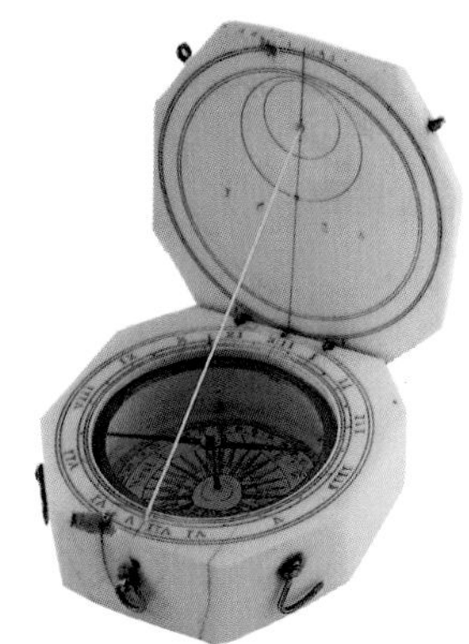

⊛ Red and black coloring. Silver gilt hooks, clasps, and hanging loop on II, silver gilt (?) hook clasp on I, copper hinges. The upper plate is considerably thinner than the bottom. The bottom of the compass bowl has a paper insert with the directions labeled in English (later addition). A thick lead ring holds the curved glass over the compass bowl in place. Compass needle is unusual and plain (probably a replacement).

Ia A concentric scale of three ranks: the outer is labeled "DL" and counts 1–15; the middle is labeled "OR" and counts 3–12 and 1–3, with the numbers 7 and 11 appearing twice; the inner scale is labeled "QI" and counts down "4 3 2 1 0" thrice.

Ib Decorated with a motif of double circles and a crescent. Two holes provided for the string gnomon. Dated 1624.

IIa Horizontal dial with a single scale for approximate latitude 37.5°, labeled V–XII–VII in red numerals, with dots marking the half-hours. Paper insert (probably a replacement) in the compass bowl labeled in English indicates a magnetic declination of 5° west of north. Insert has scale counting 0°–90° four times, beginning at north and south. Compass bowl is unusually deep.

IIb A circular table, to which the year 1625 corresponds to the number 11. Months of the year (labeled Y, F, M, A, M, Y, Y, A, S, O, N, and D) on the outer edge of the table. Double-circle motif.

Two ivory sundials at the Adler Planetarium, Chicago (Nos. W-27 and W-254, dated 1630 and 1634, respectively), are comparable, one of which has the same series of numerical scales.

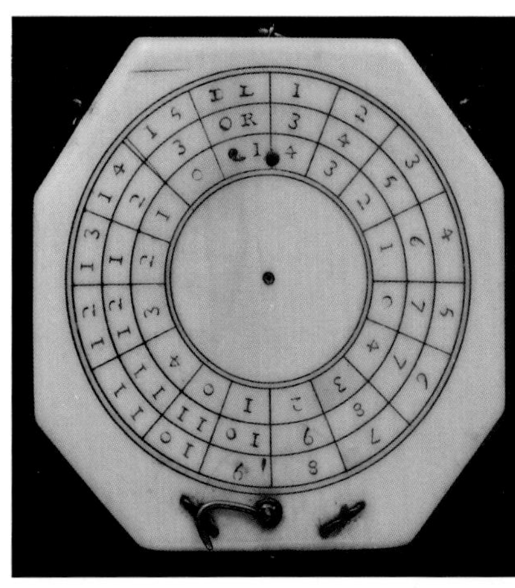

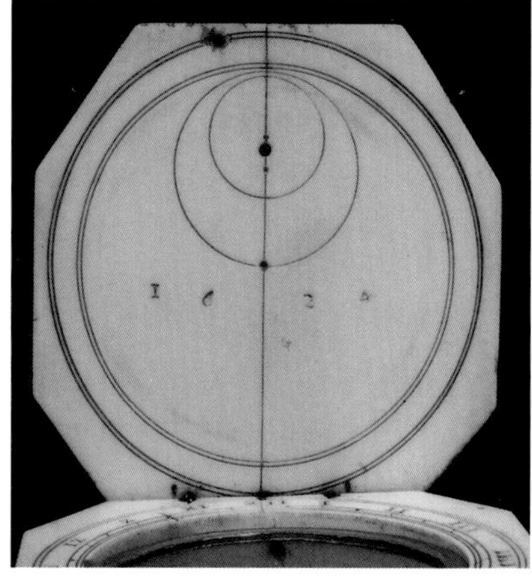

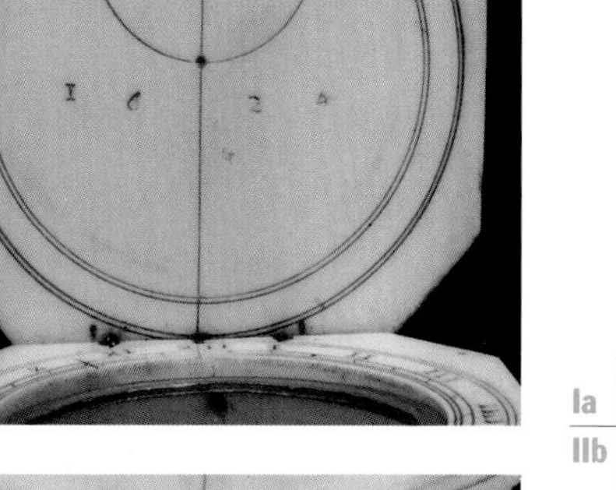

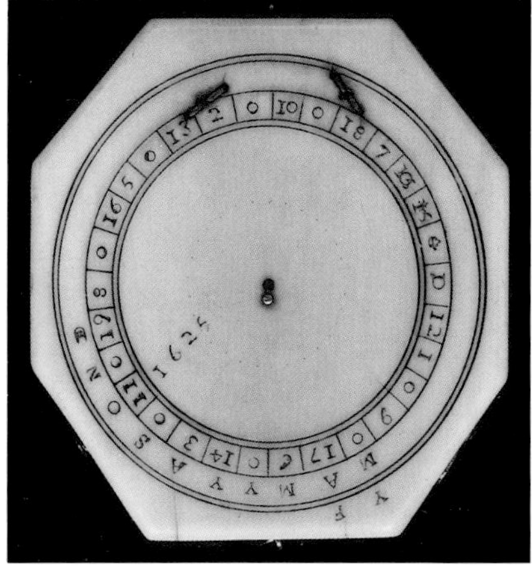

Ia | Ib
IIb | IIa

51 Rectangular Ivory Diptych

signed with maker's mark
with original case
dated 1623
Paris (?), France
6.9 (w) x 8.4 (l) x 1.4 (h) cm
case 7.7 (w) x 9.2 (l) x 2.5 (h) cm
Formerly David P. Wheatland Collection
(purchased in France, 1952)
Inventory No. 7495

⚙ Black, red, blue, and gilt coloring. Silver/copper alloy and brass clasps on I and II. Silver ring to hold compass glass intact. Glass broken. Blued-steel compass needle with gilt brass center. Ivory is slightly yellowed in places. Complete with two-piece tooled black leather case which slides open and shut. Inside edge maroon with rolled gilt ornamentation. Case has loops on either side for securing a cord so that dial could be carried.

Ia Undecorated, except for double-lined frame.

Ib Latitude table of several French cities painted in gilt on a blue background, partly disintegrated. Colorful circular painting of a fanciful castle in a mountainous landscape, surrounded by foliate ornamentation in gold on a red background. Silver spring and gilt brass sliding scale for string gnomon calibrated for 43°–48°. Brass level.

IIa Horizontal dial with three hour scales, outer scale labeled IIII–XII–VIII, middle scale unlabeled, and inner scale labeled V–XII–VII in red numerals and labeled for 43°, 45°, and 48° latitude, respectively. Dots mark the half-hours. Compass bowl labels the directions with a fleur-de-lys, ⊹, M, and O, with gilt cardinal direction lines on a blue background. Magnetic declination indicated for 10° east (in red) and 11° and 22° west (in black) of north. Brass type 2 French lunar volvelle calibrated IIII–XII–VIII, with hole through which to view the phase of the moon beneath. Blue background with gilt sun, moon, and stars at center. Double-lined edge.

IIb Undecorated, except for double-lined frame, maker's mark (possibly the letters D and N superimposed) and the date 1623. Image inverted for legibility.

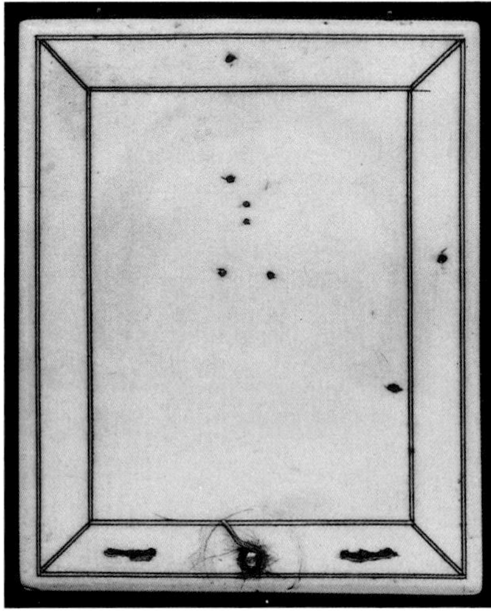

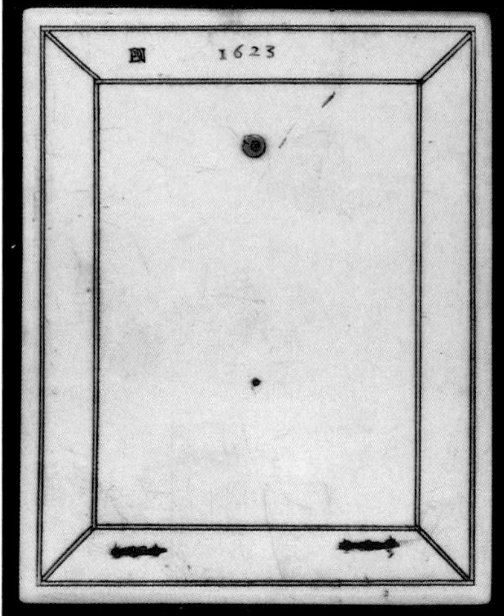

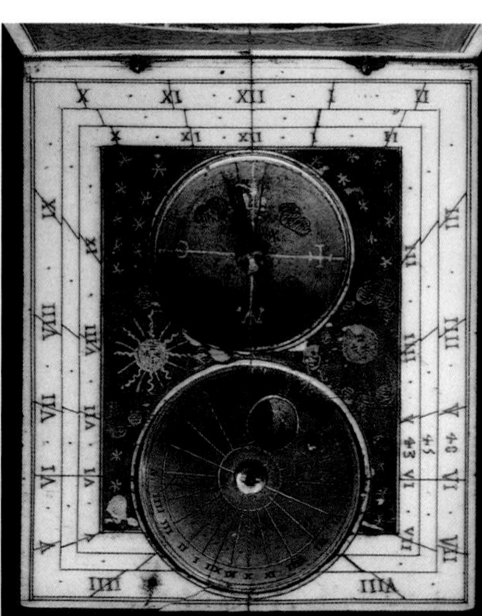

Ia | Ib

IIb | IIa

52 Rectangular Ivory Diptych

signed with maker's mark
c. 1625
Paris (?), France
5.6 (w) x 6.9 (l) x 1.5 (h) cm
Formerly David P. Wheatland Collection
(purchased in France, 1952)
Inventory No. 7499

⊛ Black, red, blue, and gilt coloring. Brass hook clasps (with one silver pin and one copper eye) and hinges on I and II. Ivory is slightly yellowed in places, especially on the rear. Typical French iron compass needle with gilt brass center. Hole in tablet II through which the string gnomon is attached.

Ia Undecorated, except for single-lined frame.

Ib Latitude table of 12 cities, most of them French (40°–50° latitude), written in gold on a round blue background, surrounded by foliate ornamentation in gold. Silver spring, brass slider, silver wire slidebar. Sliding scale for string gnomon calibrated for 40°–50°.

IIa Horizontal dial with three scales, labeled IIII–XII–VIII in the middle scale in red numerals for 42° (inner), 45° (middle), and 48° (outer), with dots marking the half-hours. Compass bowl with light blue background and gilt and gray compass arrows; eight-pointed star labels the four directions NOR, EST, SVD, and OEST. Magnetic declination 5° east of north. Glass and silver ring intact. Gilt (?) brass type 2 French lunar volvelle with viewing-hole calibrated 4–12–8 over a red and blue painted background.

IIb Undecorated, except for single-lined frame. Maker's mark (appears to be either the letters R or n followed by an I).

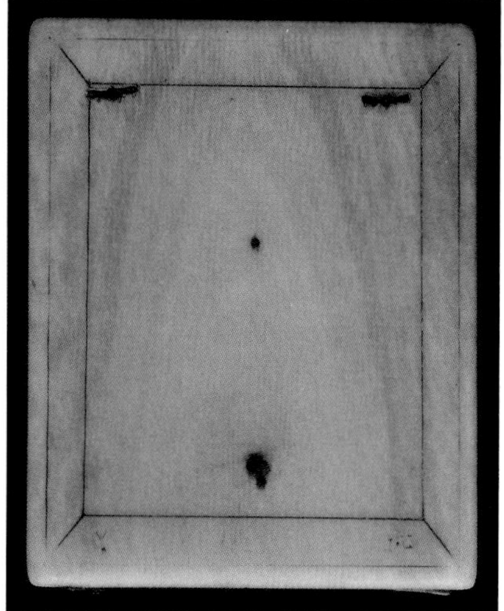

Ia | Ib
IIb | IIa

53 Rectangular Ivory Diptych

unsigned
c. 1625
Paris (?), France
5.1 (w) x 6.7 (l) x 1.4 (h) cm
Formerly Drecker Collection No. 175,
then David P. Wheatland Collection
Inventory No. 7500

⚙ Black, red, blue, and gilt coloring. Copper (?) hook clasps and eyes, brass hinges on I and II. Ivory is spotted in places. Brass plates (later addition) on top cover to hold the top plate together. Iron compass needle has a small jewel at its center in a silver mount.

Ia Double-lined edging surrounds ten-pointed geometric star motif.

Ib Latitude table of 13 cities, most of them French (latitudes 40°–50°), written in gold (now very faint) on a blue background. Geometric ornamentation surrounds latitude table. Brass spring and slider, copper wire slidebar. Sliding scale for string gnomon calibrated 40°–50°.

IIa Horizontal dial with three scales, labeled V–XII–VII in the middle scale in red numerals for 42° (inner), 45° (middle), and 48° (outer), with dots marking the half-hours. Compass bowl is painted blue and includes an eight-pointed star with faint gilt and black writing. Magnetic declination 2° east of north. Single-lined frame. Glass missing.

IIb Undecorated, except for double-lined edging. Numbered 105 in red. Several small cracks and scratches.

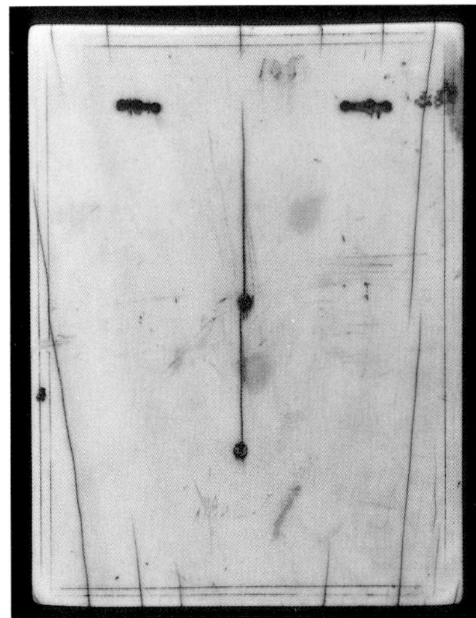

Ia | Ib
IIb | IIa

54 Rectangular Ivory Diptych

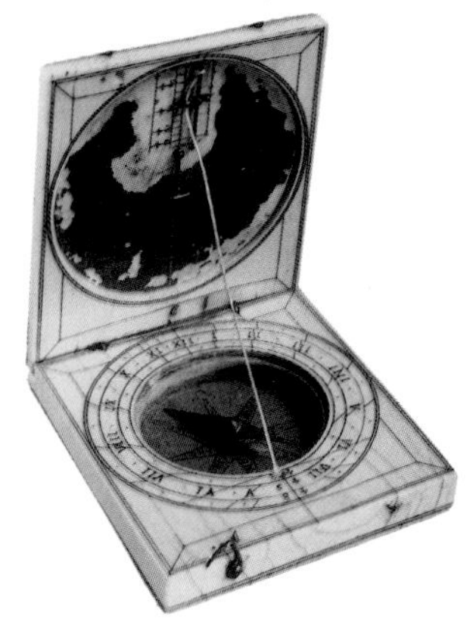

unsigned
c. 1625
Paris (?), France
6.1 (w) x 7.4 (l) x 1.7 (h) cm
Formerly Harold Gillingham Collection
No. 91, then David P. Wheatland
Collection
Inventory No. 7497

⊕ Black, blue, and gilt coloring. Brass hook clasps and hinges on I and II (right clasp on II missing), copper eyes. Ivory is slightly yellowed in places. Double-lined border is found on the outer edges of both I and II. Iron compass needle with brass center (later replacement?).

Ia Double-lined frame surrounds a three-scale German-style lunar volvelle with a brass disc (4.3 cm). Scales calibrated 1–30 (outer), 1–12 twice (middle), and 1–12 twice on the rotating brass disc in black numerals.

Ib Latitude table of 15 (?) cities, most of them French, written in gold on a round blue background, surrounded by foliate ornamentation in gold. Paint is badly deteriorated. Brass slider and slidewire, spring missing. Sliding scale for string gnomon calibrated for 40°–50°. Brass level. Double- and single-lined border makes a frame around central circular inset.

IIa Horizontal dial with three scales labeled V–XII–VII in the middle scale in black numerals for 42° (inner), 45° (middle), and 48° (outer), with dots marking the half-hours. Compass bowl has an eight-pointed star on a blue paper background with gilt and black writing. Directions labeled NOR, EST, NOR, and OEST, with other markings. Magnetic declination about 1° east of north. Double- and single-lined border is part of a frame around central compass bowl.

IIb Undecorated, except for double-lined frame.

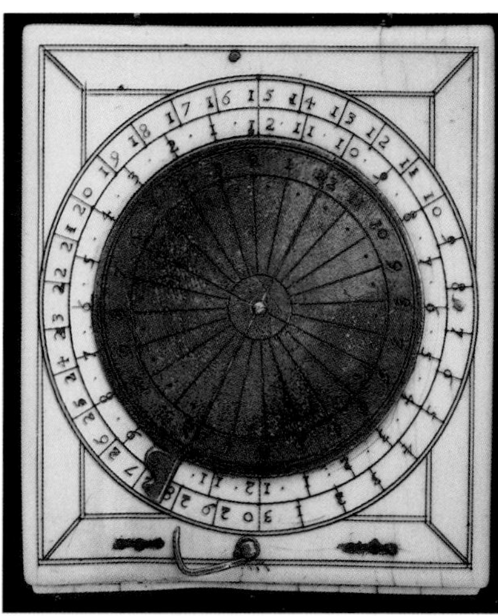

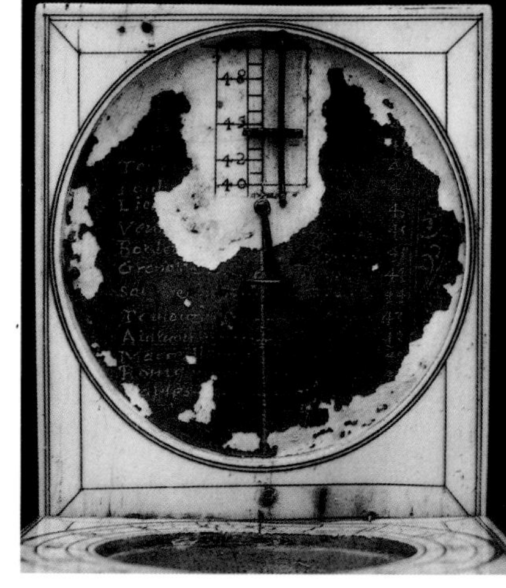

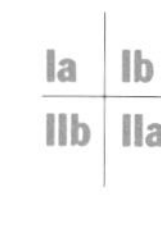

Ia	Ib
IIb	IIa

55 Rectangular Magnetic Azimuth Diptych

signed "A. André"
dated 1642
Paris, France
12.3 (w) x 14.7 (l) x 2.9 (h) cm
Formerly Chadenal Collection (sale
catalogue 24 May 1956, lot 32), then
David P. Wheatland Collection; exhibited
in "Instruments et outils d'autrefois,"
Musée des Arts Decoratifs, 1936
Inventory No. 7498

● Black and red coloring. Silver hinges
and scythe-shaped clasps on I and II.
Double-lined border on outer edges of I
and II. Two silver supports for setting the
latitude angle of the equinoctial dial on leaf
I are stored in grooves on the right side of
IIa. The longer support is used to set this
angle in autumn and winter, the shorter in
spring and summer. The lengths of the two
supports differ by the thickness of leaf I.
Three recesses in II with pivoting brass
cover-plates, one on the right side for the
weight and string (both missing) for use
with the quadrant on IIb, and two in the
front for the equinoctial dial gnomons.
Compass needle intact. The dial is one of
the largest sundials made from a single piece
of ivory.

Ia Double-lined frame and two circular
scales, both in black. The outer scale for the
days and the months (the former labeled
10–20–28, 30 or 31 and the latter labeled
with the first letter of each month), and the
inner, an equinoctial dial hour scale (for
use in spring and summer), labeled I to XII
twice with divisions every 15 minutes.
Central silvered brass rotating disc (8.3 cm,
replacement) is pivoted on a screw threaded
into the ivory. Inscribed on the disc is a
perpetual calendar square, identical to those
on Bloud-type magnetic azimuth dials, surrounded by a scale calibrated 1–30. Modern
paper label reads "71311 32 416."

Ib Latitude scale on right side for 0° to 80°
latitude to set angle of upper tablet. Border
painted red with names of three towns and
their latitudes in gold (faded). Four corners
inside the border are decorated in gold on a
dark blue background. Names and latitudes
(in degrees and minutes) of 36 cities, mostly
in France. Lunar volvelle consisting of an
upper rotating disc (replacement) with

circular and rectangular apertures, a freely
rotating ring (replacement) of larger
diameter than the disc, and three scales.
The outer scale is an hour scale (which can
also be used as an equinoctial dial for the
autumn and winter) inscribed on the ivory
I–XII twice in red numerals. The middle
scale, on the outer edge of the rotating
ring, is inscribed with a calendar labeled
10–20–28, 30 or 31 for each month from
JANVIER through DECEMBRE. There is a
small knob to turn the ring at the point
of 5 JUILLET. The inner scale, which is
inscribed on the ivory and can be read
through the rectangular aperture in the
rotating disc, is an age of the moon scale
labeled 1–30 in red numerals. Through the
circular aperture of the disc the lunar phase
is visible. The disc and ring rotate in a recess inside the hour scale. The ring rotates
freely beneath the disc which holds it in
place. The disc itself is secured by a screw
threaded into the screw in Ia.

IIa Corners painted with a red background
on which the names of 18 places and their
latitudes (in degrees and minutes) are
marked in gold. Horizontal dial has a single
scale for approximately 48.3°, and is labeled
V–XII–VII in red with divisions every 15
minutes. The compass bowl is fitted with a
silver magnetic azimuth dial surrounded by
a silver ring with a degree scale from 0 to
360 labeled every 10°. In order for this dial
to work, the magnetic azimuth dial plate in
the illustration opposite should be rotated
90° clockwise so that the 12 o'clock
position aligns with the 360° mark on the
degree scale.

IIb Horary quadrant (calibrated for just
under 49° latitude) with hour lines labeled
IIII–XII–VIII and shadow square with silver
sights that fold out for use. Months indi-

cated by zodiacal symbols. Quadrant
calibrated 5°–90° in 5° increments. Shadow
square marked in later script "Umbre
Verse" and "Umbre Dr." Underside of left
and right sights engraved respectively "A.
André" and "A Paris 1642" beneath sights.

The Harvard dial, though considerably
larger, is similar in many respects to the
instrument in the Whipple Museum,
Cambridge, England (inv. no. Wh:1704)
and described in Bryden (1988, cat. no. 90).
The Whipple dial, which is not signed,
contains a horary quadrant almost identical
to that on the Harvard dial but incorporates
a nocturnal on Ia and a simple compass in
IIa. The Adler Planetarium in Chicago has
a comparable dial (inv. no. M-257) signed
Pi(erre Duj?)ardin and dated 16(2?)7. This
dial, which is about the same size as the
Harvard instrument, has a horary quadrant
on IIb and a magnetic azimuth dial on IIa,
both similar to those on the Harvard dial.
There is some doubt that the magnetic
azimuth dials on the Harvard and Chicago
instruments are original. The compass bowl
on both these instruments may have originally contained a simple compass like that
of the Whipple dial. The circular scales on
Ia of the Harvard and Chicago instruments
are very close to those of the nocturnal on
the Whipple Museum dial. Almost certainly
the Harvard and Chicago dials originally
had the disc and index of a nocturnal on
face Ia. All three instruments have an equinoctial dial and a lunar volvelle on Ib.

The replacement lunar volvelle on the
Harvard instrument will not work. There
should be a rotating moon phase scale on
either the disc or the ring to be read against
the fixed hour scale.

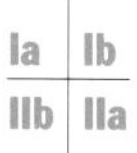
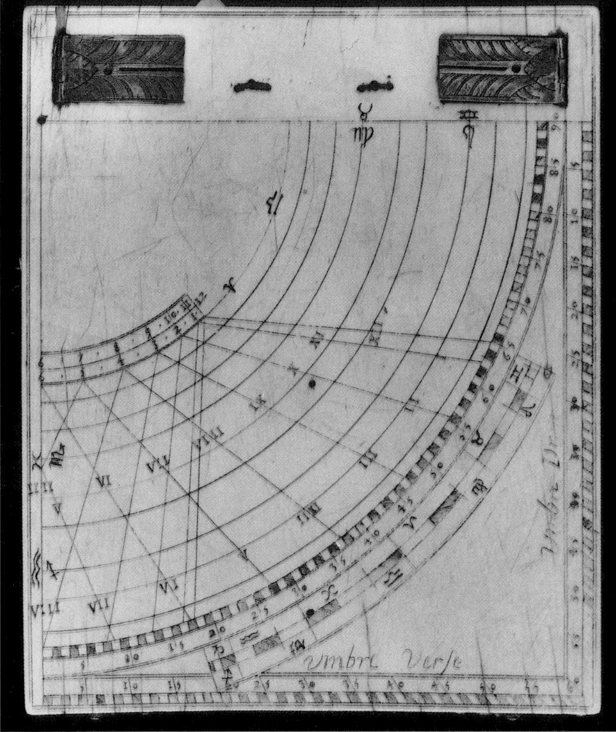

Dieppe Diptych Sundials

After about 1650 many ivory diptych sundials made in France, especially those made in Dieppe by Charles Bloud, Jacques Guerard, the Senecal workshop, and others, include a magnetic azimuth sundial set into the compass bowl in the lower tablet. The dating of these dials is difficult. One of the few (perhaps only) dated Bloud dials is cat. no. 56; its stylistic and technical sophistication indicates that Bloud was already a highly skilled craftsman by 1653. He is documented to have still been living in Dieppe in 1686. Since the only firmly dated dial is in the early 1650s, it is safe to date Bloud's productive years as the middle of the seventeenth century.

It has already been noted that there is considerable difficulty in assigning unsigned Bloud-type dials to his workshop. One stylistic link may provide a clue for those dials decorated with a poppy-flower motif (cat. nos. 67, 68, 70, and 77). Although all of these diptychs are unsigned, another dial (cat. no. 60) signed by Bloud bears this same motif. This does not guarantee that all diptychs with this motif were made by Bloud, but a case can be made at least for their manufacture in Bloud's workshop or by one of his former apprentices. In any case,

Figure 7   The elliptical hour scale on face IIa. Catalogue no. 64.

it is not unlikely that these unsigned diptychs were all manufactured by the same maker, or at least in the same workshop.

A brief mention should be made of the condition of the Bloud-type diptychs. The thin bottom of the compass bowl and typically thin upper lid appear much more liable to crack and split than the same features on contemporary German diptychs. The reason may lie both in the means of manufacture and possibly also in the source and/or quality of ivory available.

An unusual brass-pewter alloy is used on the volvelles of many Dieppe dials. The alloy has the softness of pewter and the slight yellowish tone of brass. It is likely that the pewter contains a small admixture of copper, which gives the metal a richer look yet retains the malleability of pewter (which makes the stamping or engraving of the volvelle much easier, relative to a harder surface such as silver or brass).

Azimuth dials, as their name indicates, rely upon the sun's direction to indicate the time. Because the sun's altitude varies throughout the year, these dials have to be adjusted accordingly. In the classic form of this dial as made in Dieppe, a calendar disc was incorporated into face IIb of the standard ivory dial (see figures 8 and 9). This disc, which was usually made of pewter, pivots on a brass bushing riveted to the ivory plate. The center of this bushing supports the steel pin upon which the compass needle rests. Cast into the inner surface of the calendar disc is a circular track that

Figure 8   The calendar disc on face IIb. Catalogue no. 64.

engages a lozenge-shaped guide on a small post attached to the underside of the elliptical hour scale on face IIa (see figure 7). The small post on which the guide is designed to pivot and another on the opposite side of the central bushing restrict the movement of this hour scale to a forwards/backwards motion. Therefore when the calendar disc is turned to a given date against the fixed index, usually a finger of a hand (see figure 8), the circular track engaging the guide causes the hour scale to adjust automatically for the chosen time of year. If the dial is then turned in the direction of the sun or the moon so that the shadow of the lid completely covers the lower leaf, the position of the compass needle will indicate the time at the point of its intersection with the hour scale (although in the case of a moon observation, the time would still have to be converted to solar reckoning).

Figure 9   Face IIb with calendar disc removed (top), and the inner surface of the calendar disc (bottom). Catalogue no. 64.

56 Ivory Magnetic Azimuth Diptych

signed "Charles Bloud"
dated 1653
Dieppe, France
18.2 (w) x 21.7 (l) x 2.7 (h) cm
Gift of Mrs. Irving T. Snyder
Inventory No. 7800

● Black coloring. Elaborate decoration along the outer edges. Two silver ornate scythe clasps on I, one clasp on II. Silver hinges. Silver arm in right side of IIa to set upper tablet to proper angle. Silver clasp on left side to hold gnomon rod (missing). Compass needle, glass, and ring missing.

Ia Double-lined border with typical Dieppe "pecked" ornamentation along edges and surrounding circular equinoctial dial (labeled I–XII twice). Latitudes of 24 cities worldwide arranged in a circular fashion. At the center is the coat of arms of Henri d'Orléans, the Duke of Longueville. Elaborate engraving in corners of monstrous beasts.

Ib Double-lined border with typical Dieppe "pecked" ornamentation along edges and surrounding the type 2 French lunar volvelle. Highly decorated silver volvelle (refinished), with scales labeled 1–29 ½ (inner), I–XII twice (middle), 10–20–30 twelve times for each house of the zodiac (depicted by engravings of the zodiacal animals), and 10–20–28, 30 or 31 for each month (outermost) from *Januier* through *Decembre*. Elaborate engravings of monstrous beasts in corners. Latitude scale along right side from 0° to 90° to set angle of upper tablet.

IIa Silvered moveable elliptical scale (labeled IIII–XII–VIII) for magnetic azimuth dial. Writing on parchment compass bowl insert is extremely faint. Cardinal directions labeled *Septentrion, Oriens, Mydy,* and *Occidens.* Outer scale on parchment calibrated 0°–90° four times, beginning at north and south. Parchment and elliptical hour scale rotated to an angle of 5° east of north. Double-lined border with typical Dieppe "pecked" ornamentation along

edges. More mythical, monstrous animals and fanciful foliage engraved in the corners. Compass bowl is slightly elliptical, 17.4 cm in diameter from top to bottom, 17.0 cm in diameter from side to side.

IIb Double-lined border with ornamentation along edges and surrounding the volvelle. Silver volvelle with perpetual calendar square and calendar scale by which to set the date for the magnetic azimuth elliptical hour scale inside. Index for setting the calendar scale is the pointing finger of the winged man in the upper left quadrant. The index pointer is 5° west of north to compensate for the magnetic declination of 5° east of north. Also included are scales of the epacts, dominical letters, and the dates for Easter and Pentecost from 1653 to 1700. Inscribed "Faict & Inuente par Charles Bloud A Dieppe ANNO MDCLIII." More fanciful engraving.

This dial is perhaps the largest ivory diptych ever made. It weighs 3.5 pounds (1.6 kilos)! Note that each tablet is actually made from two smaller tablets spliced together.

This is an exceptional diptych sundial for many reasons. It is one of the few French ivory diptychs that does not include a string gnomon; in other words, it has no horizontal or vertical dial, typically the hallmark of a diptych sundial. Technically, it is a relatively simple dial: it includes a magnetic azimuth dial, an equinoctial dial, and a lunar volvelle. Its elaborate decoration and its sheer size are what make this diptych so distinctive. It is also the only Dieppe magnetic azimuth dial in the collection which compensates for magnetic declination.

One technical error is apparent on the epact and liturgical calendar scale. The epact for the final year, 1700, should be 9 rather than 10, according to the Gregorian reform; apparently the maker neglected to consider the shift in the epact sequence each one or two centuries. The dominical letter underwent a similar shift at the turn of the century.

The coat of arms on Ia indicates that this dial was made for Henri d'Orléans, the Duke of Longueville (1595–1663). This illustrious personage made a name for himself in politics, but was jailed in 1650 for his activities in the Fronde. He then retired from political life to Normandy, the region in France where Dieppe is located and where Charles Bloud had his workshop. The family estate of Longueville is located just south of Dieppe.

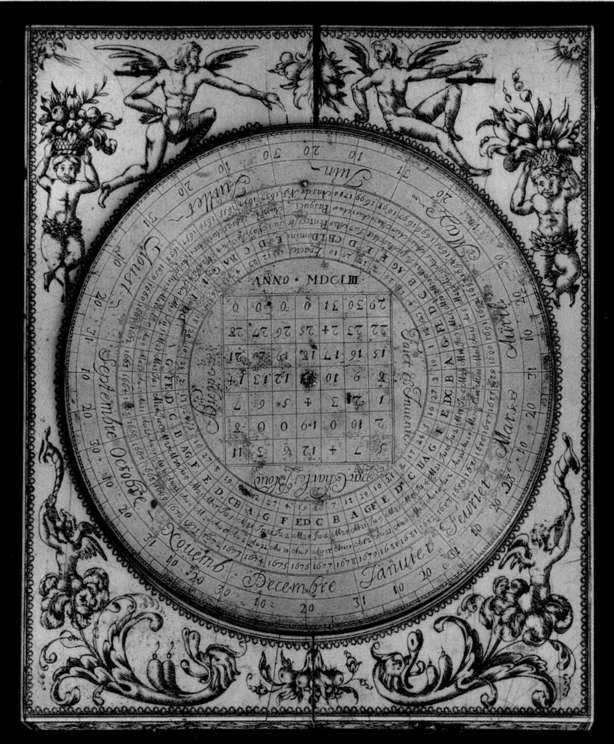

57 Ivory Magnetic Azimuth Diptych

signed "Charles Bloud"
mid to late 17th century
Dieppe, France
7.1 (w) x 7.5 (l) x 1.4 (h) cm
Formerly Ernst Collection No. 97
Inventory No. 7897

 Black coloring. Tablet II cracked at right side. Two silver scythe clasps on II. Silver clasp covers recess in left side of II for gnomon rod (missing). Silver arm in IIa to set upper tablet to proper angle. Copper wire hinges. Compass needle and silver ring intact, glass missing. Blued-steel compass needle with corroded gilt brass center. Double-lined border on outer edges of I and II. Number 97 written in pencil along hinge and on a modern paper label on Ia.

Ia Double-lined frame with typical Dieppe "pecked" ornamentation along edges, surrounding circular equinoctial dial (labeled 1–12 twice, for use in spring and summer months), and along edges of central polar dial (labeled 8–12–4). Central metal rivet with hole for pin gnomon.

Ib Double-lined border with typical Dieppe "pecked" ornamentation along edges and surrounding type 1 French lunar volvelle. Pewter volvelle with scales labeled 1–10 (inner), 1–12 twice (middle) and 10–20–28, 30 or 31 for each month (outer) from *Januier* through *Decembr,* with circular hole through which the lunar phase is visible. Floral decoration on top disc. Latitude scale along right side from 0° to 80° to set angle of upper tablet.

IIa Horizontal dial with a single hour scale for approximate latitude 47.5°, labeled 5–12–7. Silver elliptical scale (labeled V–XII–VII) for the magnetic azimuth dial. Printed paper compass bowl insert with latitudes of 16 cities, most of them French, arranged around an 8-pointed star. Outer scale on paper insert calibrated 0°–90° four times, starting at north and south. Double-lined border with typical Dieppe "pecked" ornamentation along edges and surrounding the horizontal dial hour scale.

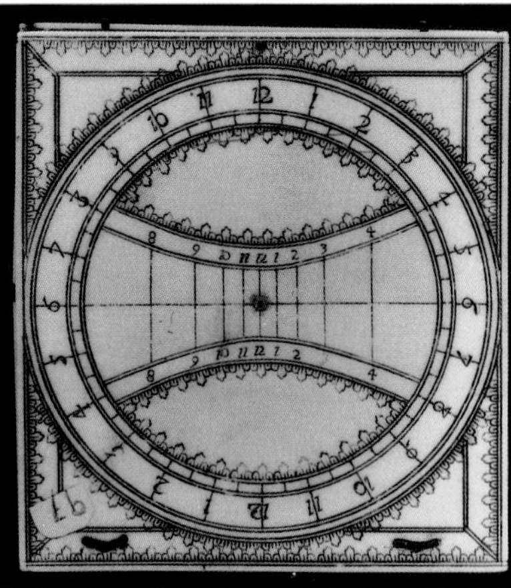

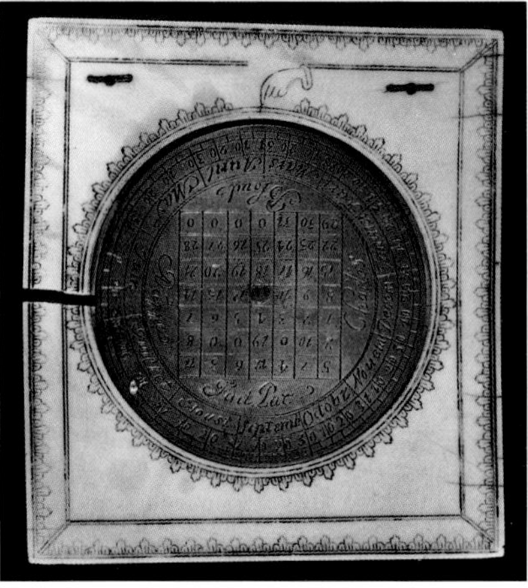

IIb Double-lined frame with typical Dieppe "pecked" ornamentation along edges and surrounding volvelle. Silver volvelle with perpetual calendar square and calendar scale to set the date for the magnetic azimuth elliptical hour scale inside. Inscribed "Fait par Charles Bloud A Dieppe" on volvelle.

58 Ivory Magnetic Azimuth Diptych

signed "Charles Bloud"
mid to late 17th century
Dieppe, France
5.8 (w) x 6.9 (l) x 1.3 (h) cm
Formerly Drecker Collection No. 159,
then David P. Wheatland Collection
Inventory No. 7513

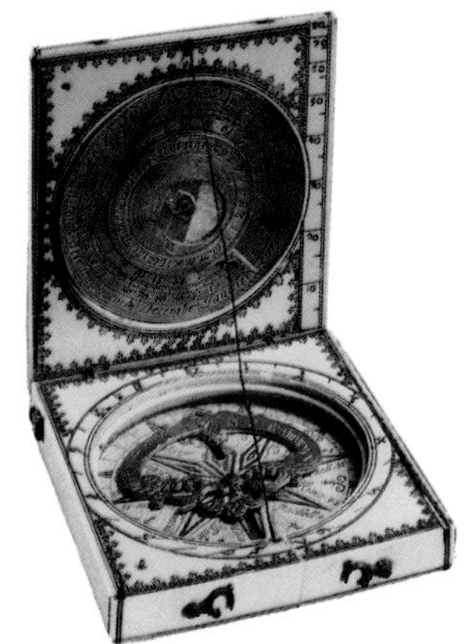

❂ Black coloring. Tablets I and II cracked, II slightly warped. Double-lined border along outer edges. Two brass scythe clasps on II. Cast brass clasp on left side of II to hold gnomon rod (missing). Brass wire hinges. Brass arm in IIa to set upper tablet to proper angle. Compass needle, glass, and ring missing. Double-lined border on outer edges of I and II.

Ia Double-lined frame with typical Dieppe "pecked" ornamentation along edges, surrounding circular equinoctial dial (labeled 1–12 twice, for use in spring and summer months), and along edges of central polar dial (labeled 8–12–4). Central metal rivet with hole for pin gnomon.

Ib Double-lined border with typical Dieppe "pecked" ornamentation along edges and surrounding type 1 French lunar volvelle. Pewter volvelle with scales labelled 1–30 (inner), 1–12 twice (middle) and 10–20–28, 30 or 31 for each month (outer) from *Januier* through *Decembr,* with circular hole through which the lunar phase is visible (top disc missing). Latitude scale along right side from 0° to 80° to set angle of upper tablet.

IIa Horizontal dial with a single hour scale for approximate latitude 46.5°, labeled 5–12–7. Pewter elliptical scale (labeled V–XII–VII) for the magnetic azimuth dial. Printed paper compass bowl with insert with latitudes of 14 cities, most of them French (42°–52° latitude), arranged around an 8-pointed star. Double-lined border with typical Dieppe "pecked" ornamentation along edges.

IIb Double-lined frame with typical Dieppe "pecked" ornamentation along edges and surrounding volvelle. Edge along hinge slightly beveled. Pewter volvelle with per-

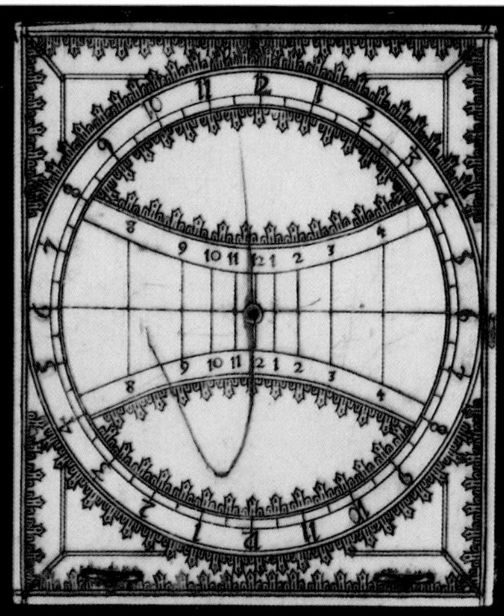

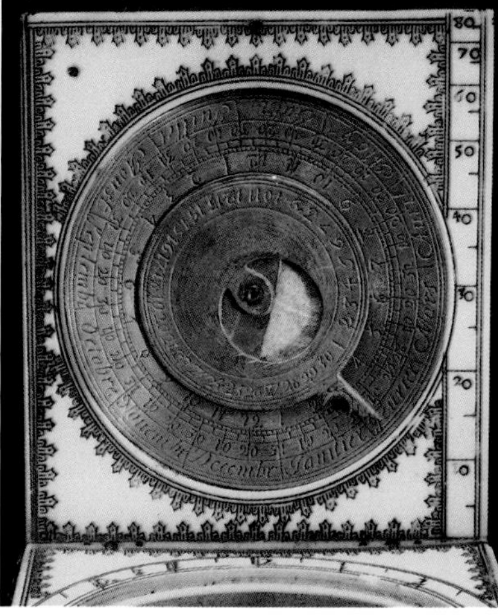

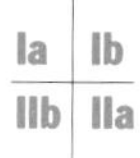

petual calendar square and calendar scale to set the date for the magnetic azimuth elliptical hour scale inside. Inscribed "Fait & inuen par Charles Bloud A Dieppe" on volvelle. Numbered 38 in red ink.

59 Ivory Magnetic Azimuth Diptych

signed "Charles Bloud"
mid to late 17th century
Dieppe, France
8.7 (w) x 10.2 (l) x 1.7 (h) cm
Formerly Drecker Collection No. 157,
then David P. Wheatland Collection
Inventory No. 7514

⚫ Black coloring. Tablets I and II cracked
lengthwise. Two pewter scythe clasps on II
(added later). Hole in left side of II to hold
gnomon rod (missing). Clasp also missing.
Brass arm in IIa to set upper tablet to proper
angle. Brass wire hinges. Blued-steel compass needle (probably a replacement), glass
(replacement), and metal ring intact.
Double-lined border on outer edges of I
and II.

Ia Elaborate double-lined border with typical Dieppe "pecked" ornamentation along
edges, surrounding circular equinoctial dial
(labeled I–XII twice, for use in spring and
summer months), and along edges of central
polar dial (labeled VIII–XII–IIII). Central
metal rivet with hole for pin gnomon.

Ib Triple-lined border with typical Dieppe
"pecked" ornamentation along edges and
surrounding type 1 French lunar volvelle.
Decorated pewter volvelle with scales
labeled 1–30 (inner), I–XII twice (middle),
and 10–20–28, 30 or 31 for each month
(outer) from *Januier* through *Decembr,* with
circular hole through which the lunar phase
is visible. Latitude scale along right side
from 0° to 80° to set angle of upper tablet.

IIa Horizontal dial with a single hour scale
for approximate latitude 46°, labeled V–XII–
VII. Elliptical scale missing. Printed paper
compass bowl insert with latitudes of 14
cities, most of them French (42°–52° latitude), arranged around an 8-pointed star.
Middle scale on paper insert calibrated
0°–90° four times, starting at north and
south, outer scale (added later?) labeled
1–24. Triple-lined border with typical
Dieppe "pecked" ornamentation along
edges and surrounding compass bowl.

IIb Elaborate double-lined border with
ornamentation along edges. Pewter-brass

Ia | Ib
IIb | IIa

alloy volvelle with perpetual calendar
square and calendar scale to set the date for
the magnetic azimuth elliptical hour scale
inside. Inscribed "Fait par Charles Bloud A
Dieppe" on volvelle. Numbered 152 in
red ink.

60 Ivory Magnetic Azimuth Diptych

signed "Charles Bloud"
mid to late 17th century
Dieppe, France
8.6 (w) x 10.9 (l) x 1.7 (h) cm
Formerly David P. Wheatland Collection
(purchased from F. A. E. Staack, 1959)
Inventory No. 7519

❂ Octagonal diptych. Black coloring. Tablet I slightly curved. Two silver scythe clasps on II. Hole for gnomon rod (missing) on left side of II. Clasp also missing. Silver arm on right side of IIa to set upper tablet to proper angle. Copper wire hinges. Compass needle intact, glass and ring missing. Blued-steel compass needle with gilt brass center. Elaborate decoration along outer edges. Modern paper label along right edge reads "W 259."

Ia Ornate double-lined border with Dieppe "pecked" ornamentation along edges, surrounding circular equinoctial dial (labeled I–XII twice, for use in spring and summer months), and along edges of central polar dial (8–12–4). Inscribed "C. BLOVD A. DIEPPE." Poppy flower motif in regions between the equinoctial and polar dials. Central metal rivet with hole for pin gnomon.

Ib Ornate double-lined border with Dieppe "pecked" ornamentation along edges and surrounding type 1 French lunar volvelle. Nicely engraved pewter volvelle with scales labeled 1–30 (inner), I–XII twice (middle), and 10–20–28, 30 or 31 for each month (outer) from *Januie* through *Decemb,* with circular hole through which the lunar phase is visible. Floral motif on central disc. Latitude scale along right side from 0° to 80° to set angle of upper tablet.

IIa Horizontal dial with a single hour scale for approximate latitude 49°, labeled V–XII–VII, with half-, quarter-, and eigth-hours marked. Pewter elliptical scale (labeled V–XII–VII) for the magnetic azimuth dial. Printed paper compass bowl insert (painted yellow, red, and blue) with latitudes of 24 cities, most of them French (40°–53° latitude). Cardinal directions labeled *Septentrion, Orient, Mydy,* and

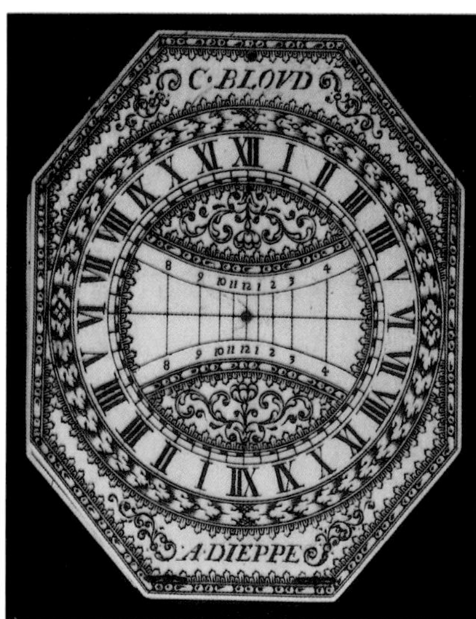

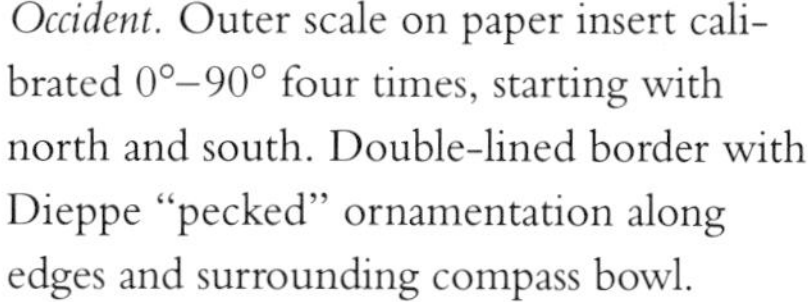

Ia | Ib
IIb | IIa

Occident. Outer scale on paper insert calibrated 0°–90° four times, starting with north and south. Double-lined border with Dieppe "pecked" ornamentation along edges and surrounding compass bowl.

IIb Very ornate double-lined border with typical Dieppe "pecked" ornamentation along edges and surrounding volvelle. Silver

volvelle with perpetual calendar square and calendar scale to set the date for the magnetic azimuth elliptical hour scale inside. Inscribed "Fait par Charles Bloud A Dieppe" on volvelle.

61 Ivory Magnetic Azimuth Diptych

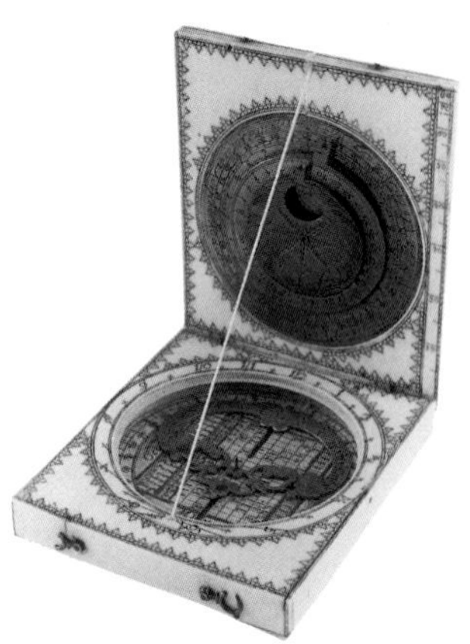

signed "Jacques Senecal"
mid to late 17th century
Dieppe, France
6.8 (w) x 8.0 (l) x 1.5 (h) cm
Formerly David P. Wheatland Collection
(purchased in France, 1952)
Inventory No. 7510

⚫ Black coloring. Tablet I split lengthwise
in the middle. Brass arm at right side of IIa
to set angle for equinoctial dial. Two pewter
scythe clasps (one broken) on II. Copper
wire hinges. Compass needle and glass
missing. Cast pewter clasp on left side of II
to hold gnomon rod (missing). Double-
lined border on outer edges of I and II.

Ia Double-lined border with distinctive
Dieppe "pecked" ornamentation along
edges and surrounding circular equinoctial
dial labeled 1–12 twice, for use in spring
and summer months. Central metal rivet
with hole for pin gnomon.

Ib Double-lined border with distinctive
Dieppe "pecked" ornamentation along
edges and surrounding type 1 French lunar
volvelle. Moveable brass volvelle with scales
labeled 1–30 (inner), 1–12 twice (middle),
and 10–20–28, 30 or 31 for the months
(outer) from *Januier* through *decembre,* with
circular hole through which the lunar phase
is visible. Central volvelle has an 8-pointed
star and cloud motif. Latitude scale along
right side from 0° to 80° to set angle for
upper tablet.

IIa Horizontal dial with a single hour
scale for approximate latitude 47°, labeled
5–12–7. Printed paper insert (painted red,
yellow, and green) in compass bowl with
gazeteer of the attributes of 18 French cities.
Cardinal directions labeled *Septentrion,
Orient, Mydy,* and *Occident.* Outer scale on
insert calibrated 0°–90° four times, beginning at north and south. Pewter-brass alloy
elliptical magnetic azimuth dial scale labeled
V–XII–VII. Double-lined border with distinctive Dieppe "pecked" ornamentation
along edges and surrounding the horizontal
dial hour scale.

IIb Double-lined border with distinctive
Dieppe "pecked" ornamentation along the
edges and surrounding the volvelle. Brass
volvelle with perpetual calendar square and
calendar scale to set the date for the magnetic azimuth elliptical hour scale inside.
Inscribed "Jacques Senecal A Dieppe fecit"
on the volvelle.

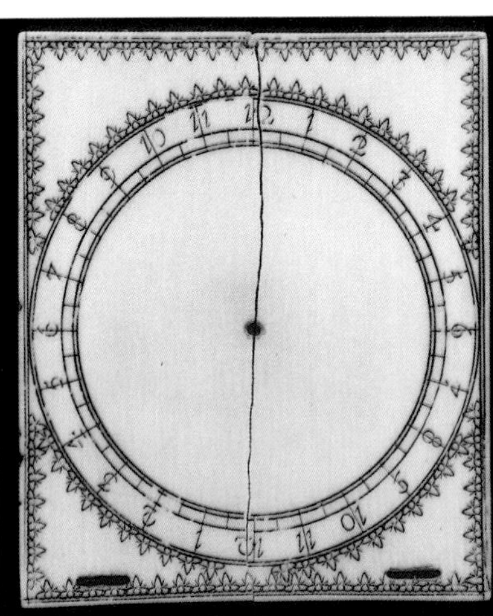

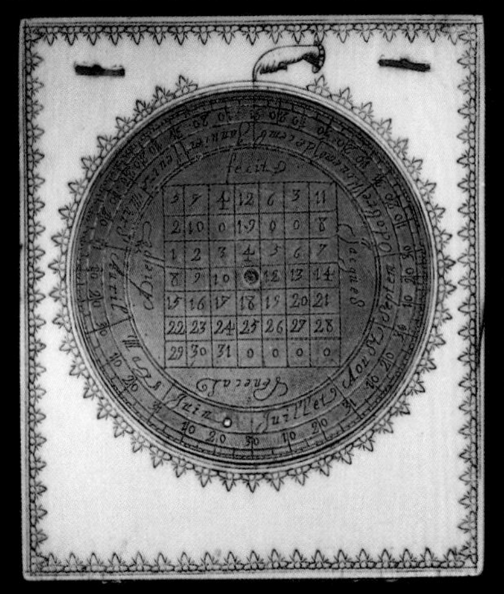

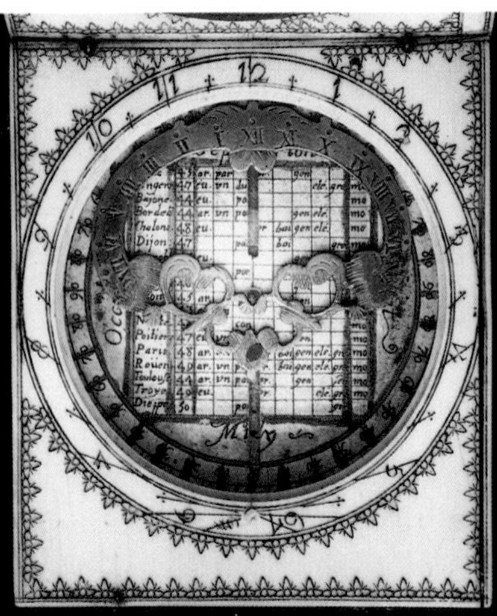

<table>
<tr><td>Ia</td><td>Ib</td></tr>
<tr><td>IIb</td><td>IIa</td></tr>
</table>

62 Ivory Magnetic Azimuth Diptych

signed "Jacques Guerard"
mid to late 17th century
Dieppe, France
6.5 (w) x 7.3 (l) x 1.6 (h) cm
Formerly Harold Gillingham Collection
No. 100, then David P. Wheatland
Collection
Inventory No. 7507

⚫ Black coloring. Brass arm at right side
of IIa to set angle for equinoctial dial. Two
pewter scythe clasps on II. Blued-steel
compass needle with gilded brass center.
Double-lined border on outer edges of I
and II. Glass is a replacement. Pewter clasp
on left side of II to hold gnomon rod
(missing).

Ia Double-lined border with typical Dieppe
"pecked" ornamentation along edges and
surrounding circular equinoctial dial labeled
1–12 twice, for use in spring and summer
months. Central metal rivet with hole for
pin gnomon.

Ib Double-lined border with typical Dieppe
"pecked" ornamentation along edges and
surrounding type 1 French lunar volvelle.
Ornate brass moveable volvelle labeled
1–30 (inner scale), 1–12 twice (middle), and
10–20–28, 30 or 31 for each month (outer
scale) from *Januier* through *decembre,* with
circular hole through which the lunar phase
is visible. Pink fabric beneath the volvelle.
Latitude scale along right side from 0° to
80° to set angle for upper tablet.

IIa Horizontal dial with a single hour
scale for approximate latitude 45°, labeled
4–12–8 on a brass insert on top of compass
bowl glass. Printed paper insert in compass
bowl with a gridded gazeteer of the attributes of 18 French cities. Cardinal directions labeled *Septentrion, Orient, Mydy,* and
Occident. Pewter-brass alloy (?) elliptical
magnetic azimuth dial hour scale labeled
5–12–7. Outer scale calibrated 0°–90° four
times, starting with north and south.
Double-lined border with typical Dieppe
"pecked" ornamentation along edges and
surrounding compass bowl.

IIb Double-lined border with typical
Dieppe ornamentation along the edges and

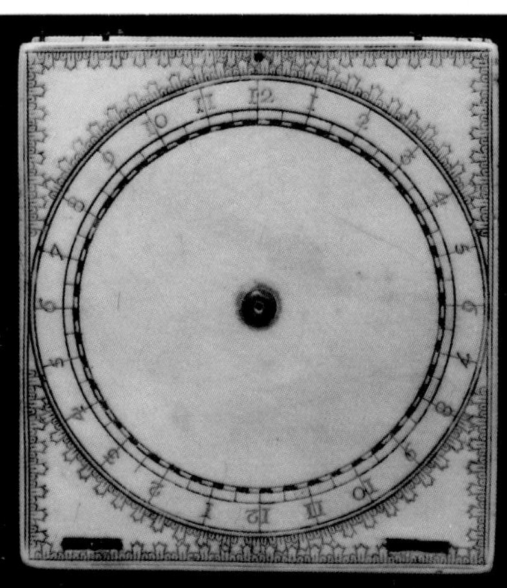

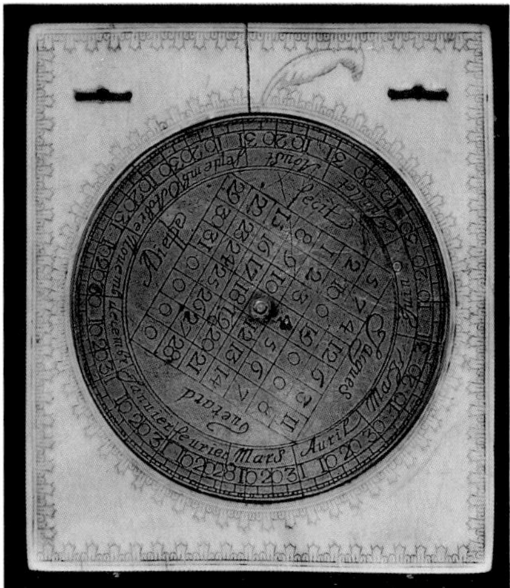

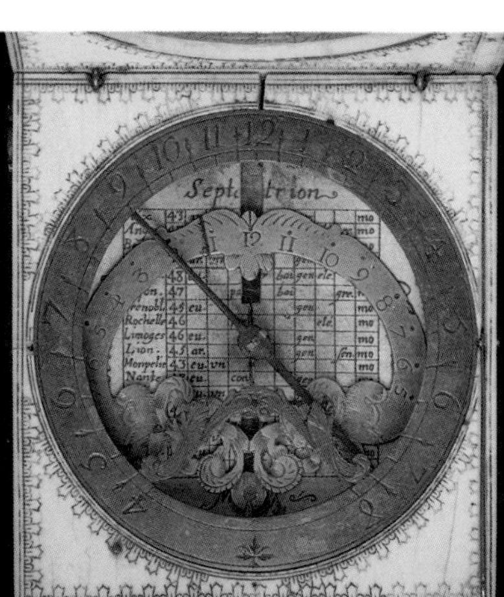

surrounding volvelle. Pewter alloy volvelle
with perpetual calendar square and calendar
scale to set the magnetic azimuth elliptical
hour scale inside. Inscribed "Jacques
Guerard A Dieppe fecit" on volvelle.

Ia	Ib
IIb	IIa

63 Ivory Magnetic Azimuth Diptych

signed "Jacques guerard"
mid to late 17th century
Dieppe, France
6.4 (w) x 7.3 (l) x 1.6 (h) cm
Formerly Harold Gillingham Collection
No. 101, then David P. Wheatland
Collection
Inventory No. 7508

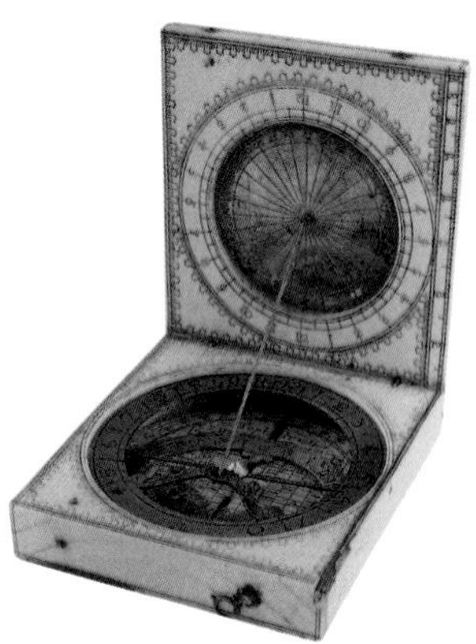

⚫ Black coloring. Brass arm at right side
of IIa to set angle for equinoctial dial.
Single pewter-brass alloy scythe clasp on II.
Copper wire hinges. Blued-steel compass
needle with gilt brass center. Glass is a
replacement. Cigar-shaped brass gnomon
rod in recess on left side of II, held in place
by a wooden peg. Clasp missing. Double-
lined border on outer edges of I and II.

Ia Double-lined border with typical Dieppe
"pecked" ornamentation along edges and
surrounding circular equinoctial dial labeled
1–12 twice, for use in spring and summer
months. Central metal rivet with hole for
pin gnomon.

Ib Double-lined border with typical Dieppe
"pecked" ornamentation along edges and
surrounding type 2 French lunar volvelle.
Pewter-brass alloy moveable volvelle labeled
1–30 (inner scale) and 1–12 twice (outer
scale, on ivory, also to be used as equinoctial dial scale for use in autumn and winter
months). Latitude scale along right side
from 0° to 80° to set angle for upper tablet.

IIa Horizontal dial with a single hour scale
for approximate latitude 37°, labeled
4–12–8 on a brass insert on top of compass
bowl glass. Printed paper insert in compass
bowl with gridded gazeteer of the attributes
of 18 French cities. Brass elliptical magnetic
azimuth dial scale in compass bowl labeled
V–XII–VII. Cardinal directions labeled
Septentrion, Orient, Mydy, and *Occident.*
Double-lined border with typical Dieppe
"pecked" ornamentation along edges and
surrounding compass bowl.

IIb Double-lined border with typical
Dieppe "pecked" ornamentation along the
edges and surrounding the volvelle. Pewter-
brass alloy volvelle with perpetual calendar
square and calendar scale to set the date for

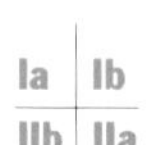

the magnetic azimuth elliptical scale inside.
Inscribed "Jacques guerard A Dieppe fecit"
on volvelle. Corner apparently burned.

64 Ivory Magnetic Azimuth Diptych

unsigned
mid to late 17th century
Dieppe, France
6.1 (w) x 6.8 (l) x 1.3 (h) cm
Formerly Harold Gillingham Collection
No. 99b, then David P. Wheatland
Collection
Inventory No. 7389

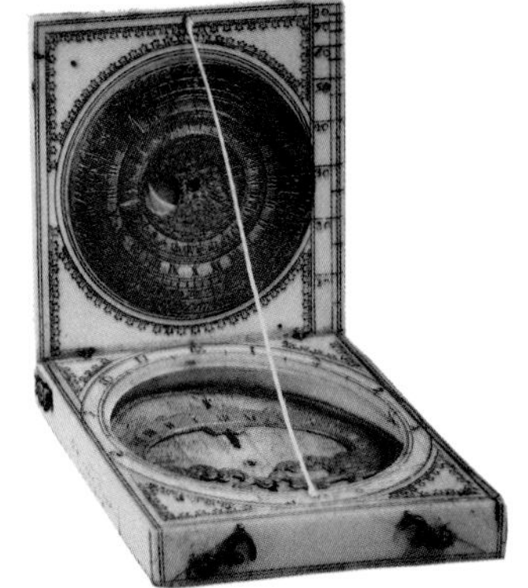

la | lb
llb | lla

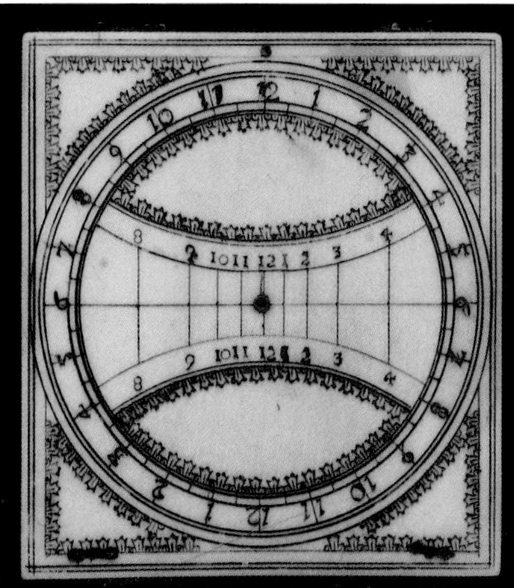

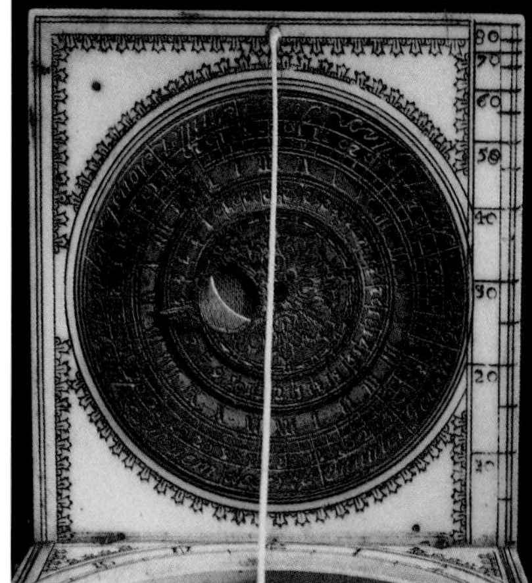

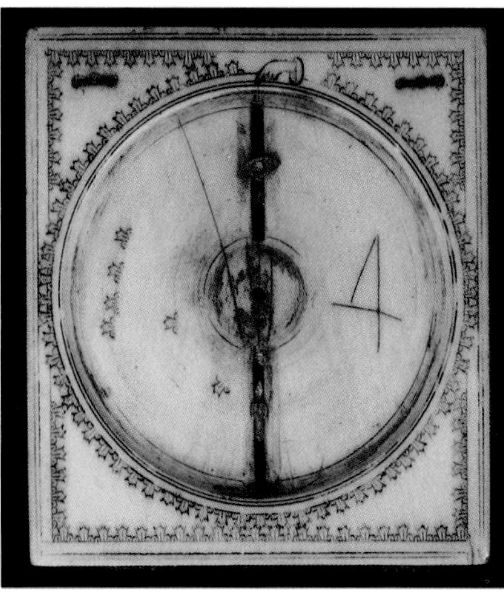

⚙ Black coloring. The calendar disc on
IIb is unattached. Two brass scythe clasps
on left side of II. Pewter clasp on left side
of II to hold gnomon rod (missing). Brass
arm at right side of IIa to set angle for
equinoctial dial. Compass needle and glass
missing. Double-lined border along outer
edges. Modern paper label along left edge
reads "C. Bloud. Dieppe. 1660-1680." in
black ink. Labeled 389 in black ink along
right edge.

Ia Double- and single-lined border with
typical Dieppe "pecked" ornamentation
along edges, surrounding circular equinoctial dial (labeled 1–12 twice, for use in
spring and summer months), and at edges
of central polar dial (labeled 8–12–4). Central metal rivet with hole for pin gnomon.

Ib Double- and single-lined border with
typical Dieppe "pecked" ornamentation
along edges and surrounding type 1 French
lunar volvelle. Pewter volvelle with scales
labeled 1–30 (inner), I–XII twice (middle),
and 10–20–28, 30 or 31 for each month
(outer) from *Januier* through *decemb,* with
circular volvelle through which the lunar
phase is visible. Foliate decoration on top
disc. Latitude scale along right side from 0°
to 80° to set angle for upper tablet.

IIa Horizontal dial with a single hour
scale for approximate latitude 46°, labeled
5–12–7. Pewter elliptical scale for the magnetic azimuth dial labeled V–XII–VII. Very
yellowed printed paper compass bowl insert
with gazeteer of 18 French cities (41°–50°
latitude). Cardinal directions labeled
Septentrion, Orient, Mydy, and *Occident.*
Outer degree scale cut off. Double- and
single-lined border with typical Dieppe
"pecked" ornamentation along edges and
surrounding the horizontal dial hour scale.

IIb Double- and single-lined border with
typical Dieppe "pecked" ornamentation
along edges. Pewter volvelle (not shown,
see figure 8, page 117) badly tarnished,
with perpetual calendar square and calendar scale to set the date for the magnetic
azimuth elliptical hour scale inside. The
number 4 is scratched onto the rear of the
volvelle, as well as onto the ivory.

65 Ivory Magnetic Azimuth Diptych

unsigned
mid to late 17th century
Dieppe, France
7.2 (w) x 8.8 (l) x 1.6 (h) cm
Formerly Drecker Collection No. 158,
then David P. Wheatland Collection
Inventory No. 7388

⚙ Black coloring. The calendar disc on IIb is unattached. Tablet I is cracked lengthwise in several places. Two brass scythe clasps on II. Brass clasp on left side of II to hold gnomon rod (missing). Later brass clasp and eye on right side of dial to hold it open. Brass wire hinges. Brass arm at right side of IIa to set angle for equinoctial dial. Compass needle and original glass missing. Double-lined border on outer edges of I and II.

Ia Double- and single-lined border with Dieppe "pecked" ornamentation along edges, surrounding circular equinoctial dial (labeled 1–12 twice, for use in spring and summer months), and along edges of central polar dial (labeled 8–12–4). Central metal rivet with hole for pin gnomon.

Ib Double- and single-lined border with Dieppe "pecked" ornamentation along edges and surrounding type 1 French lunar volvelle. Very badly corroded pewter volvelle with scales labeled 1–30 (inner), I–XII twice (middle), and 10–20–28, 30 or 31 for each month (outer) from *Januier* through *Decembr,* with circular hole through which the lunar phase is visible. Foliate decoration on top disc. Volvelle is somewhat bent and does not fit its recess properly. Latitude scale along right side from 0° to 80° to set angle for upper tablet.

IIa Horizontal dial with a single hour scale for approximate latitude 48°, labeled 5–12–7. Badly corroded pewter elliptical scale (labeled V–XII–VII) for the magnetic azimuth dial (scale upside-down in the illustration). Printed paper compass bowl insert with gazeteer of 18 cities, most of them French. Cardinal directions labeled *Septentrion, Orient, Mydy,* and *Occident.* Outer scale on paper insert calibrated

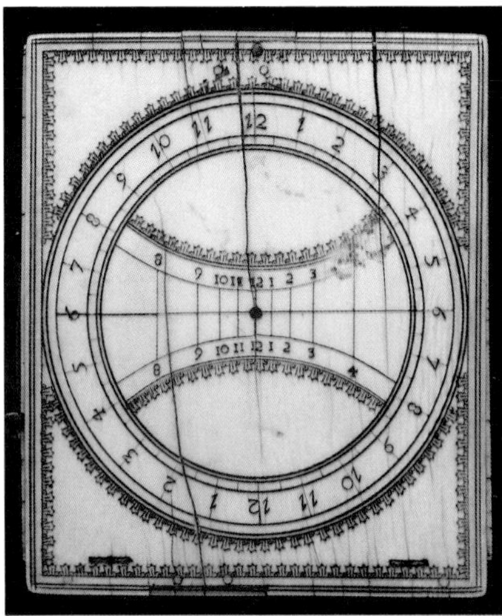

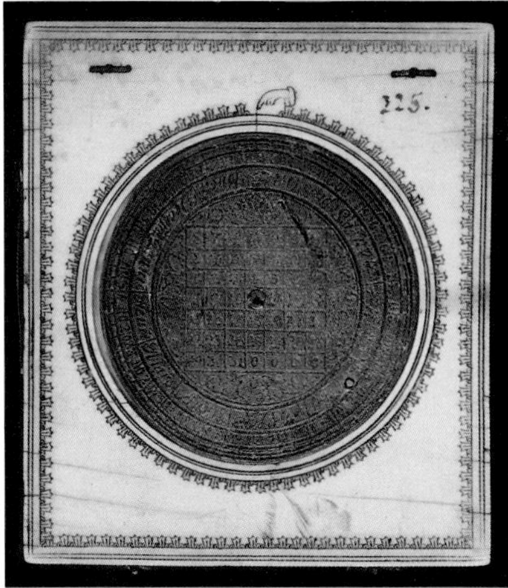

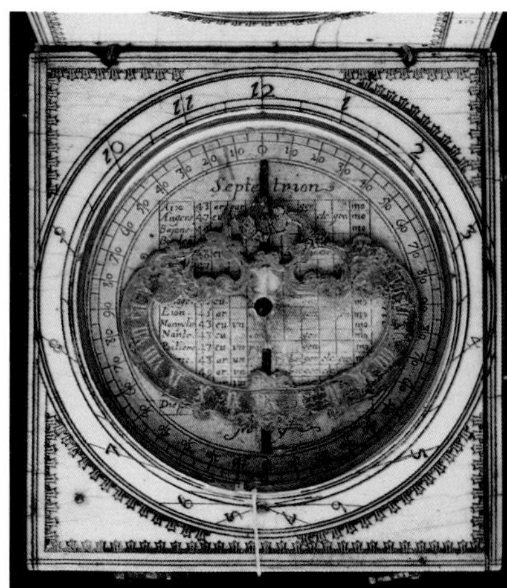

Ia | Ib
IIb | IIa

0°–90° four times, starting at north and south. Double- and single-lined border with "pecked" ornamentation along edges and surrounding most of the horizontal dial hour scale.

IIb Double- and single-lined border with typical Dieppe "pecked" ornamentation along edges and surrounding volvelle. Pewter volvelle is corroded, with perpetual calendar square and calendar scale to set the

magnetic azimuth elliptical hour scale inside. Modern paper label on rear of volvelle reads "dr 158." Number 225 written in black ink on ivory.

66 Ivory Magnetic Azimuth Diptych

unsigned
mid to late 17th century
Dieppe, France
5.8 (w) x 7.3 (l) x 1.3 (h) cm
Formerly Harold Gillingham Collection
No. 92, then David P. Wheatland
Collection
Inventory No. 7518

⚙ Octagonal diptych. Black coloring. Tablet II cracked lengthwise. Two brass scythe clasps on II. Brass clasp (added later) on left side of IIa to hold gnomon rod (missing). Brass wire hinges. Pewter-brass alloy arm in IIa to set upper tablet to proper angle. Compass needle and original glass missing, silver ring intact. Double-lined border on edges of I and II. Modern paper label on right edge reads "Campbell 500. C.Bloud 1660/80 Dieppe" in black ink.

Ia Double-lined border with typical Dieppe "pecked" ornamentation along edges, surrounding circular equinoctial dial (labeled 1–12 twice, for use in spring and summer months), and along edges of central polar dial (labeled 8–12–4). Central metal rivet with hole for pin gnomon.

Ib Double-lined border with typical Dieppe "pecked" ornamentation along edges and surrounding type 1 French lunar volvelle. Pewter volvelle with scales labeled 1–30 (inner), 1–12 twice (middle), and 10–20–28, 30 or 31 for each month (outer) from *Januier* through *Decembr,* with circular hole through which the lunar phase is visible. Foliate decoration on top disc. Latitude scale along right side from 0° to 80° to set angle of upper tablet.

IIa Horizontal dial with a single hour scale for approximate latitude 47.5°, labeled 5–12–7. Brass elliptical scale (labeled 5–12–7) for the magnetic azimuth dial. Badly darkened printed paper compass bowl insert with latitudes of several cities, mostly French (latitudes 40°–50°), arranged around an 8-pointed star. Black line indicates a magnetic declination of 26° west of north. Outer scale on paper insert calibrated 0°–90° four times, starting at north and south. Double-lined border with typical Dieppe

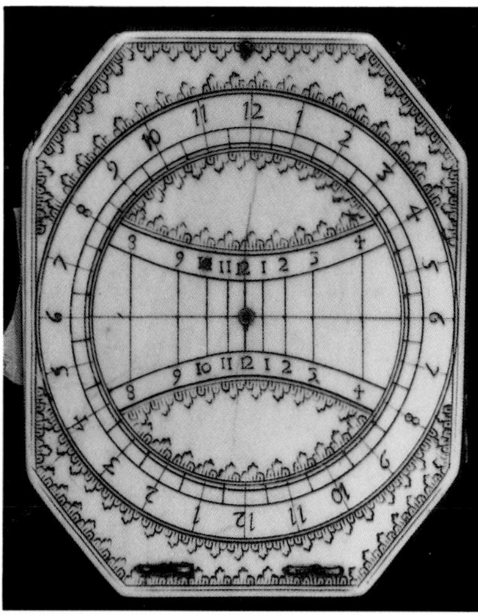

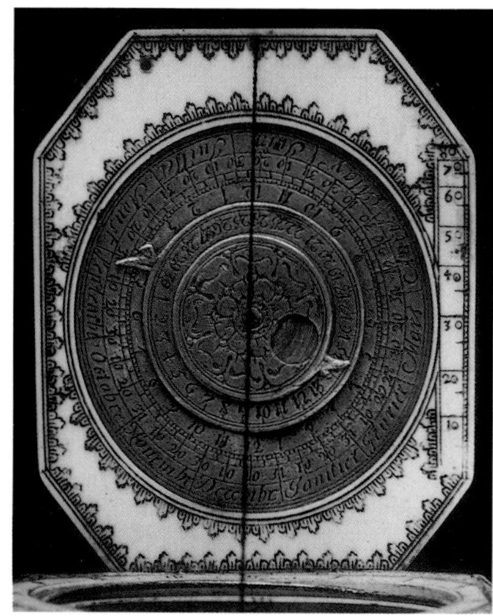

Ia | Ib
IIb | IIa

"pecked" ornamentation along edges and surrounding horizontal dial hour scale.

IIb Double-lined border with typical Dieppe "pecked" ornamentation on edges and surrounding volvelle. Pewter-brass alloy volvelle with perpetual calendar and calendar scale to set the date for the magnetic azimuth elliptical hour scale inside.

67 Ivory Magnetic Azimuth Diptych

unsigned
mid to late 17th century
Dieppe, France
8.2 (w) x 9.5 (l) x 1.5 (h) cm
Formerly David P. Wheatland Collection
Inventory No. 7517

⚫ Black coloring. This diptych was poorly restored. The thin base of the compass bowl was replaced by a disc of fiberboard, and the silvered brass parts may have been resilvered but not reassembled. Brass scythe clasps and hinges. Double-lined border on outer edges of I and II. Brass-clasped hole in left side of II for gnomon rod (missing). Brass arm at right side of IIa to set angle for equinoctial dial.

Ia Double-lined border with typical Dieppe "pecked" ornamentation and poppy-flower motif in corners and twice inside circular equinoctial dial labeled 1–12 twice, for use in spring and summer months; central polar dial labeled 7–12–5.

Ib Double-lined border with typical Dieppe "pecked" ornamentation and poppy-flower motif in corners surrounds silver type 1 French lunar volvelle. Volvelle scales labeled 1–30 (inner), 1–12 twice (middle), and 10–20–30 for each calendar month (outer) labeled *Januier* through *decembre*, with circular hole through which the lunar phase is visible. Two inner discs of volvelle not permanently attached. Latitude scale along right side from 0° to 80° to set angle of upper tablet. Three holes for setting string gnomon for latitudes 45°, 46°, and 48.5°.

IIa Horizontal dial with a single hour scale labeled 4–12–8. Double-lined border with poppy-flower motif in corners. Silver elliptical hour scale labeled 7–12–5 is all that remains in the compass bowl.

IIb Double-lined border with typical Dieppe "pecked" ornamentation and poppy-flower motif in four corners. Silvered metal disc at center with perpetual calendar square and calendar scale to set the date for the magnetic azimuth hour scale inside.

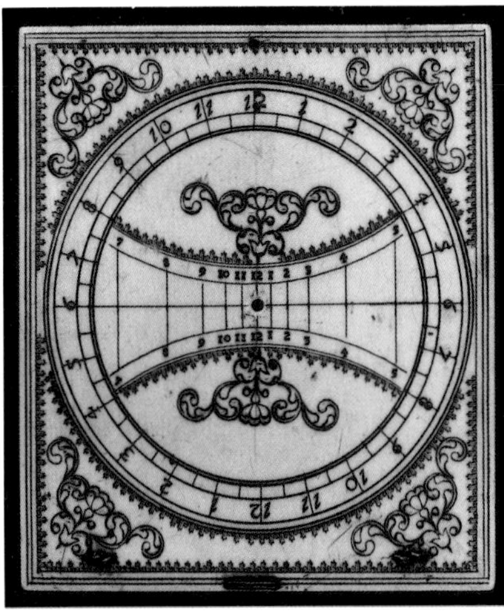

Ia | Ib
IIb | IIa

68 Ivory Magnetic Azimuth Diptych

unsigned
mid to late 17th century
Dieppe, France
7.2 (w) x 7.3 (l) x 1.2 (h) cm
Formerly Harold Gillingham Collection
No. 97, then David P. Wheatland
Collection
Inventory No. 7390

⚙ Black coloring. Tablet II is split side-ways. Brass scythe clasps and silver hinges. Double-lined border on outer edges of I and II. Brass-clasped hole in left side of II for gnomon rod (missing). Silver arm at right side of IIa to set angle for equinoctial dial. Pewter volvelle parts are not attached.

Ia Double-lined border with poppy-flower motif in four corners and twice inside circular equinoctial dial labeled 1–12 twice, for use in spring and summer months; central polar dial labeled 8–12–4. Central metal rivet with hole for pin gnomon.

Ib Double-lined frame with poppy-flower motif in four corners surrounds warped pewter type 1 French lunar volvelle. Volvelle scales labeled 1–10 (inner), 1–12 twice (middle), and 10–20–28, 30 or 31 for each month (outer scale) labeled *Januier* through *Decembr*, with circular hole through which the lunar phase is visible. Latitude scale along right side from 0° to 80° to set angle for upper tablet.

IIa Horizontal dial with a single hour scale for approximate latitude 46.5°, labeled 5–12–7. Double-lined border with poppy-flower motif in four corners. Pewter elliptical hour scale labeled V–XII–VII is all that remains of the contents of the compass bowl.

IIb Double-lined border with poppy-flower motif in four corners. Badly tarnished pewter disc at center with perpetual calendar square and calendar scale to set the date for the magnetic azimuth hour scale inside.

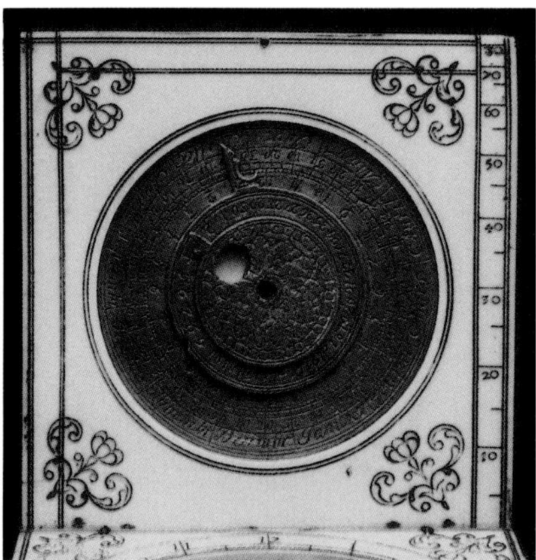

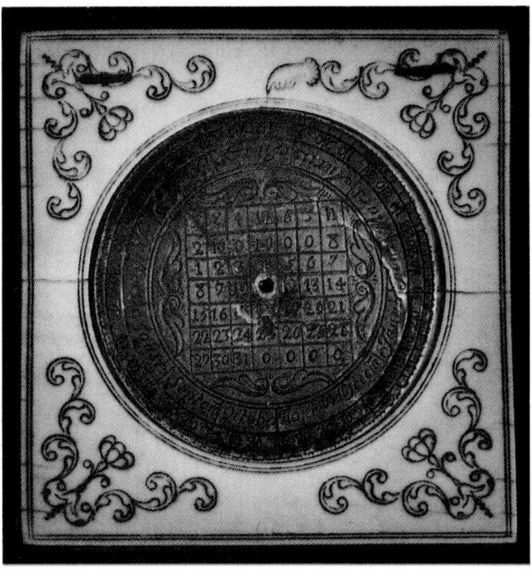

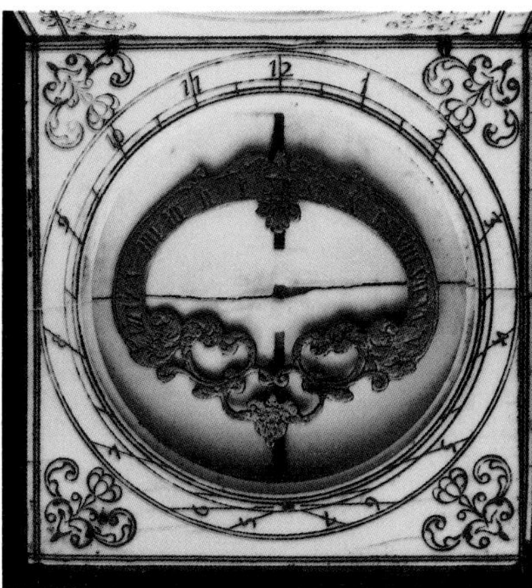

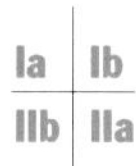

69 Ivory Magnetic Azimuth Diptych

unsigned
mid to late 17th century
Dieppe, France
6.2 (w) x 6.9 (l) x 1.3 (h) cm
Formerly Harold Gillingham Collection
No. 95, then David P. Wheatland
Collection
Inventory No. 7387

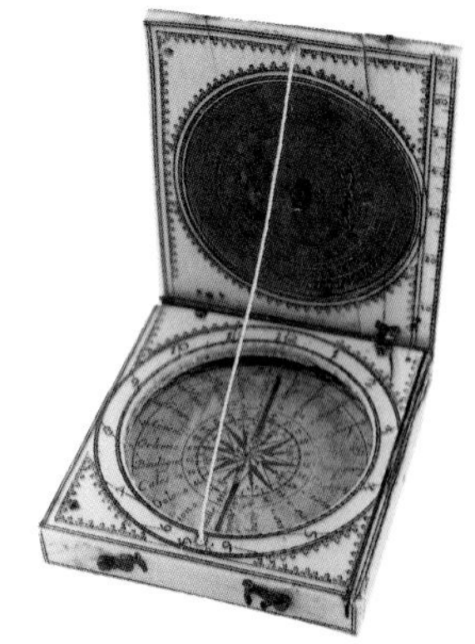

⚙ Black coloring. This dial was apparently refitted with a long brass hinge. Tablet I is cracked lengthwise and is badly warped. Hole for gnomon rod (missing) on left side of II. Clasp also missing. Brass arm at right side of IIa to set angle for equinoctial dial. Two brass scythe clasps on II, one of which is badly corroded. Compass needle and glass missing. Double-lined border on outer edges of I and II.

Ia Double- and single-lined border with typical Dieppe "pecked" ornamentation along edges and surrounding circular equinoctial dial (labeled 1–12 twice, for use in spring and summer months) and at edges of the polar dial at center (labeled 8–12–4).

Ib Double- and single-lined border with typical Dieppe "pecked" ornamentation along edges and surrounding type 1 French lunar volvelle. Very badly corroded pewter volvelle, central disc missing. Volvelle scales labeled I–XII twice (inner) and 10–20–28, 30 or 31 for each month (outer) from *Januier* through *Decembr.* Latitude scale along right side from 0° to 80° degrees to set angle for upper tablet.

IIa Horizontal dial with a single hour scale for approximate latitude 46.5°, labeled 5–12–7. Printed paper insert in compass bowl with latitude table of 24 cities, mostly French (43°–51° latitude), arranged in a circular fashion. Outer scale on insert labeled 1–12 twice. Double-lined border with typical Dieppe "pecked" ornamentation along edges and surrounding the horizontal dial hour scale.

IIb Double-lined border with typical Dieppe "pecked" ornamentation along edges and surrounding recess for volvelle (missing). Traces of paper glued to the volvelle recess remain.

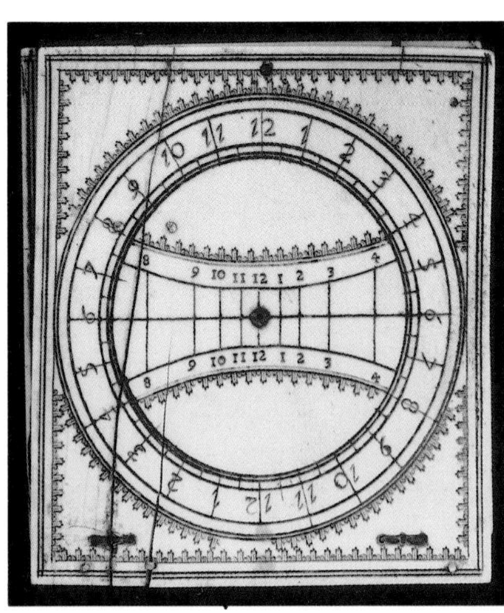

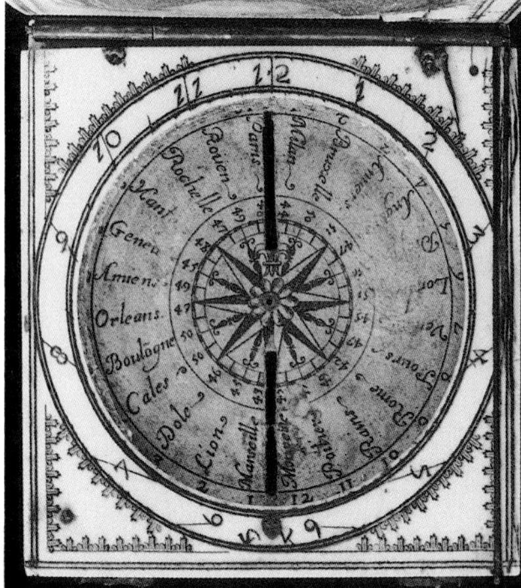

<table>
<tr><td>Ia</td><td>Ib</td></tr>
<tr><td>IIb</td><td>IIa</td></tr>
</table>

70 Ivory Magnetic Azimuth Diptych

unsigned
mid to late 17th century
Dieppe, France
6.5 (w) x 7.8 (l) x 1.5 (h) cm
Formerly David P. Wheatland Collection
(purchased from E. Weil, 1954)
Inventory No. 7386

⬤ Very faded black coloring. Brass arm at right side of IIa to set angle for equinoctial dial. All other metal parts except the brass hinges and part of one silver clasp are missing. A metal gnomon rod appears to be permanently wedged in its recess in the left side of II. Tablet I slightly warped. Numbered 386 in black ink along right outer edge. Faint double-lined border on outer edges of I and II.

Ia Double- and single-lined border with swirling ornamentation along edges. Circular equinoctial dial labeled 1–12 twice for use in spring and summer months; central polar dial labeled 8–12–4. Typical Dieppe "pecked" ornamentation along edges of polar dial. Central metal rivet with hole for pin gnomon.

Ib Double- and single-lined border with simple poppy-flower ornamentation in four corners. Double- and single-lined circles surround recess for volvelle (missing). Latitude scale along right side from 0° to 80° to set angle for upper tablet.

IIa Horizontal dial with a single hour scale for approximate latitude 46.5°, labeled 5–12–7. Double- and single-lined border with simple poppy-flower motif in four corners. Compass bowl is empty.

IIb Double- and single-lined border with ornamentation along the edges. Double- and single-lined circles surround recess for volvelle (missing).

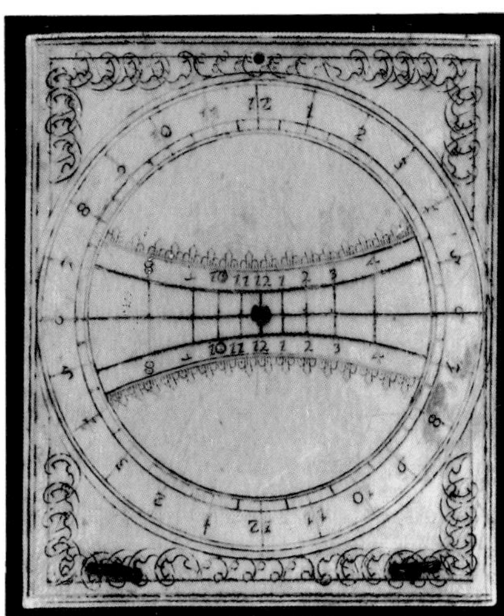

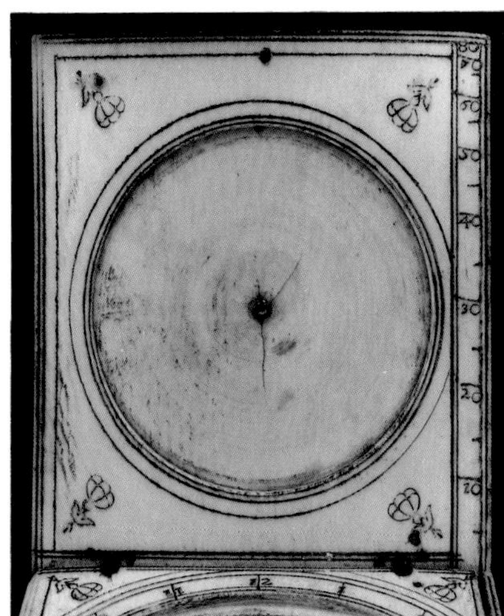

71 Ivory Magnetic Azimuth Diptych

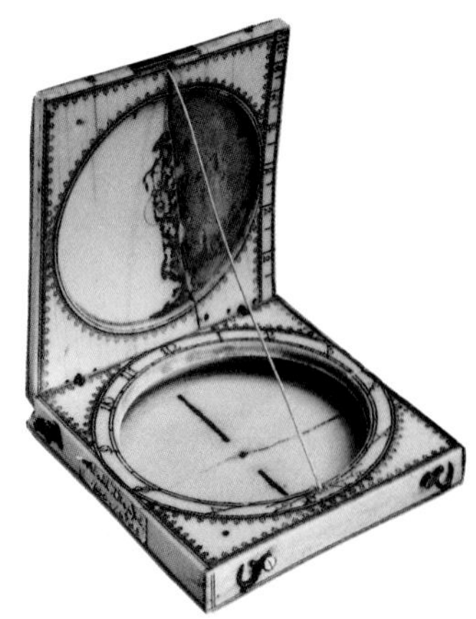

unsigned
mid to late 17th century
Dieppe, France
6.2 (w) x 6.9 (l) x 1.5 (h) cm
Formerly Harold Gillingham Collection
No. 99a, then David P. Wheatland
Collection
Inventory No. 7385

⚙ Black coloring. Tablet I is cracked lengthwise and is fitted with two small silver supports along the edges. Tablet II is split sideways, but still intact. Outer face of II is beveled along edges. Brass arm at right side of IIa to set angle for equinoctial dial. Two pewter scythe clasps on II. Brass pins. Pewter clasp on side of II to hold silver gnomon rod (a replacement). Copper wire hinges. Compass needle, glass, and all other metal parts missing. Double-lined border on outer edges of I and II. Modern paper label along left edge reads "Bloud. Dieppe. 1660-1680." in black ink.

Ia Double-lined border with typical Dieppe "pecked" ornamentation along edges, surrounding the circular equinoctial dial (labeled 1–12 twice, for use in spring and summer months), and at the edge of the central polar dial (labeled 8–12–4).

Ib Double-lined border with typical Dieppe "pecked" ornamentation along edges. Central volvelle missing. Remnants of an adhesive remain in the volvelle recess. Latitude scale along right side from 0° to 80° to set angle for upper tablet.

IIa Horizontal dial with a single hour scale for approximate latitude 47.5°, labeled 5–12–7. Compass bowl is empty. Double-lined border with typical Dieppe "pecked" ornamentation along edges and surrounding the horizontal dial hour scale.

IIb Double-lined border with typical Dieppe "pecked" ornamentation along beveled edges and surrounding volvelle recess. Volvelle missing. Mark which resembles an elongated letter Z scratched onto volvelle recess.

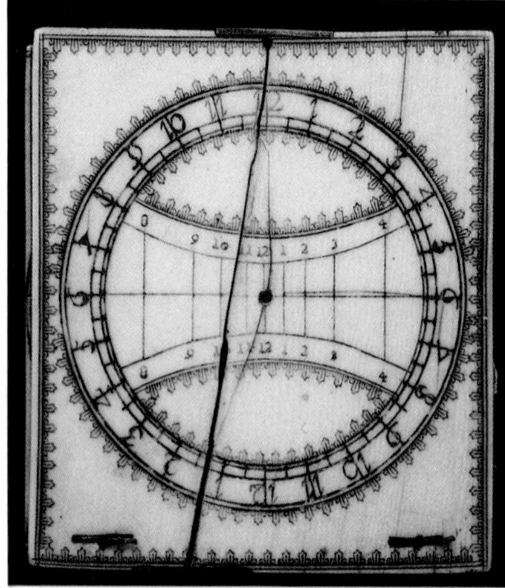

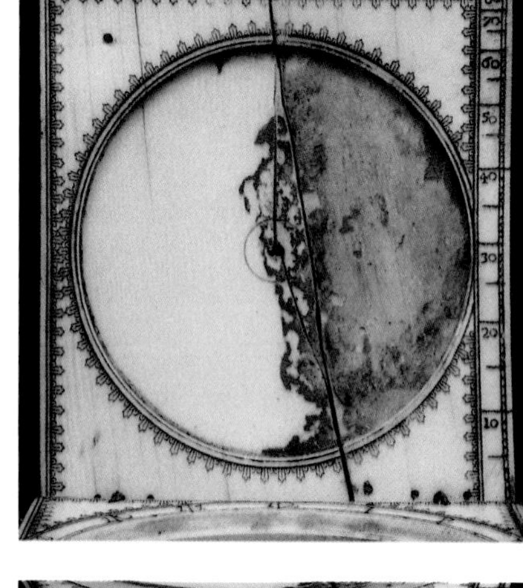

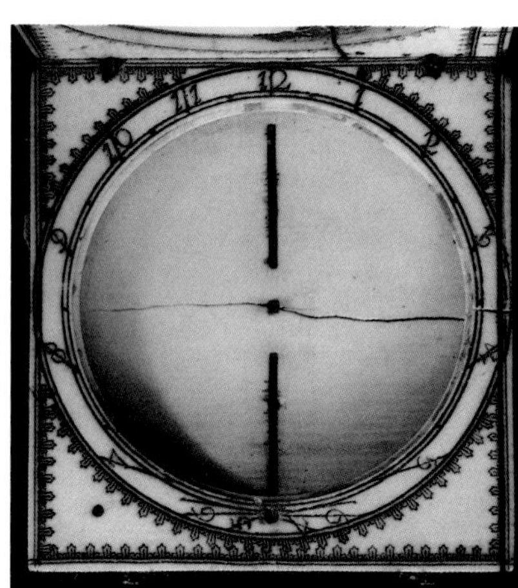

Ia | Ib
IIb | IIa

72 Rectangular Ivory Diptych

unsigned
mid to late 17th century
Dieppe, France
4.9 (w) x 5.6 (l) x 1.0 (h) cm
Formerly Harold Gillingham Collection
No. 94, then David P. Wheatland
Collection
Inventory No. 7503

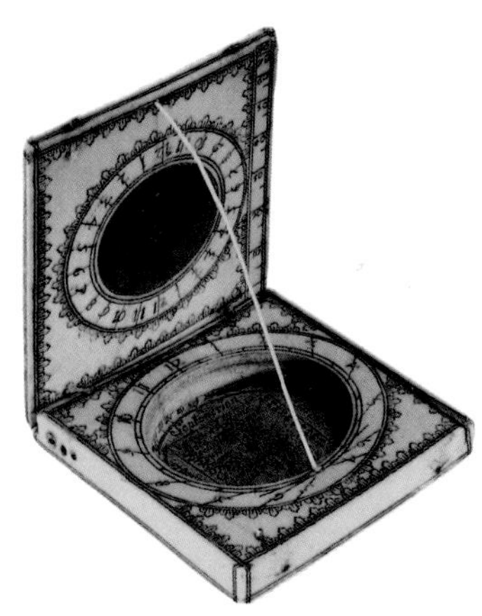

⬤ Black coloring. String hinges bind the two plates (now separated); remnants of silver hinges in II. Brass pins. Double-lined border on the outer edges of I and II. Both plates are slightly curved. Hole in left side of II for pin gnomon (missing). Clasp also missing. Brass arm at right side of IIa to set upper plate to the proper angle for equinoctial and polar dials. Compass needle and glass missing.

Ia Double-lined border with Dieppe "pecked" ornamentation surrounds a circular equinoctial dial labeled 1–12 twice (for use in spring and summer months) and the central polar dial labeled 9–3.

Ib Double-lined border with typical Dieppe "pecked" ornamentation surrounds pewter type 2 French lunar volvelle calibrated 1–30 (heavily tarnished) and another scale calibrated 1–12 twice (also to be used as scale for equinoctial dial during autumn and winter months). Latitude scale along right side from 0° to 80° to set angle of upper tablet.

IIa Horizontal dial with a single scale for approximate latitude 49.5°, numbered 5–12–7 in black numerals. Double-lined border with typical Dieppe "pecked" ornamentation. Paper latitude list of 22 cities, mostly French (42°–51° latitude), inside compass bowl. Directions labeled *Septentrion, Orient, Mydy,* and *Occident.* Outer scale on paper insert calibrated 0°–90° four times, beginning at north and south.

IIb Plain, except for double-lined frame with Dieppe "pecked" ornamentation.

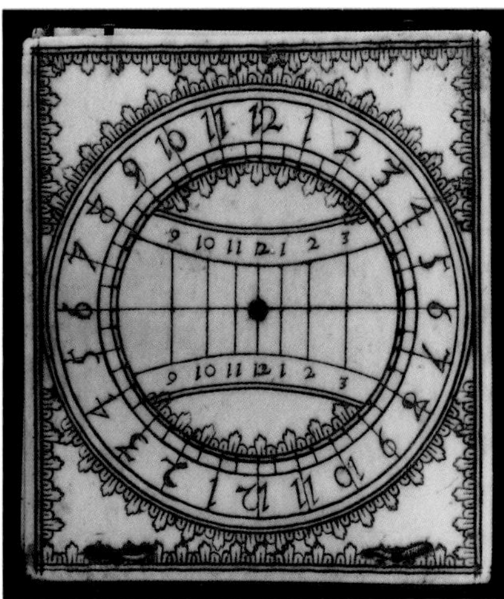

Ia | Ib
IIb | IIa

73 Rectangular Ivory Diptych

unsigned
mid to late 17th century
Dieppe, France
5.9 (w) x 6.7 (l) x 1.3 (h) cm
Formerly Harold Gillingham Collection
No. 93, then David P. Wheatland
Collection
Inventory No. 7516

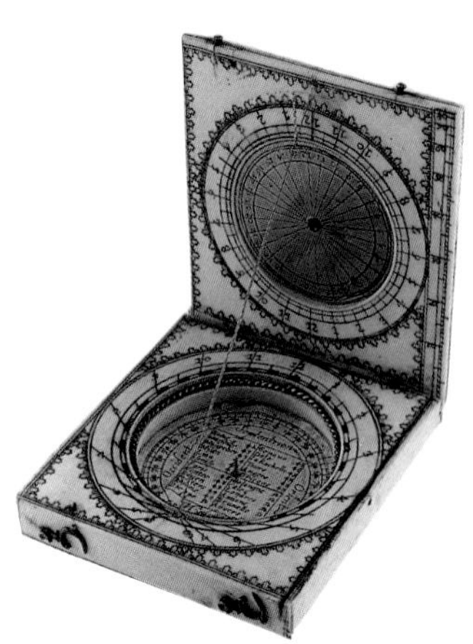

⬤ Black coloring. Silver scythe clasps
and silver/copper alloy hinges. Double-
lined border on the outer edges of I and II.
Clasped hole in left side of II for pin
gnomon (missing). Silver arm at right side
of IIa to set upper plate to the proper angle
for equinoctial dial. Compass needle and
glass missing. Braided silver ring to hold
compass glass intact. Modern label on left
side of II reads "C. Bloud. — Dieppe.
1660–1680. 3." Tablet I slightly curved.

Ia Double-lined border with typical Dieppe
"pecked" ornamentation surrounds a circular equinoctial dial labeled 1–12 twice for
use in spring and summer months. Central
metal rivet with hole for pin gnomon.

Ib Double-lined border with typical Dieppe
ornamentation surrounds silver gilt type 2
French lunar volvelle calibrated 1–30.
Outer scale calibrated 1–12 twice, also to
be used as scale for equinoctial dial during
autumn and winter months. Latitude scale
along right side from 0° to 80° to set angle
for equinoctial dial.

IIa Horizontal dial with a single scale for
approximate latitude 49°, labeled 4–12–8.
Double-lined border with typical Dieppe
"pecked" ornamentation. Printed paper
latitude list of 24 cities, mostly French
(42°–51° latitude), inside compass bowl.
Directions labeled *Septentrion, Orient, Mydy,*
and *Occident.* Outer scale on paper insert
calibrated 0°–90° four times, beginning
at north and south.

IIb Plain, except for double-lined border
with Dieppe "pecked" ornamentation.

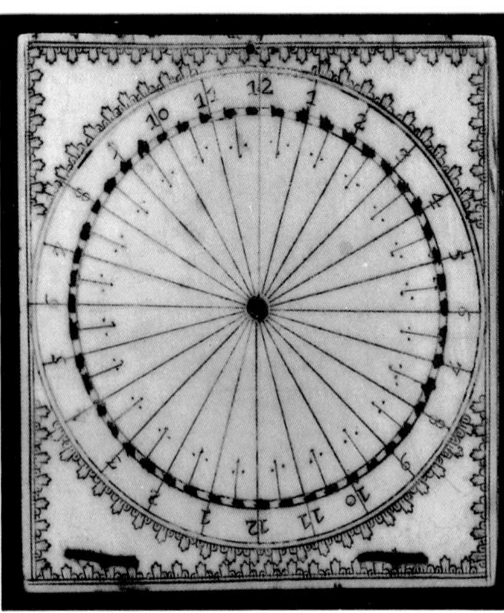

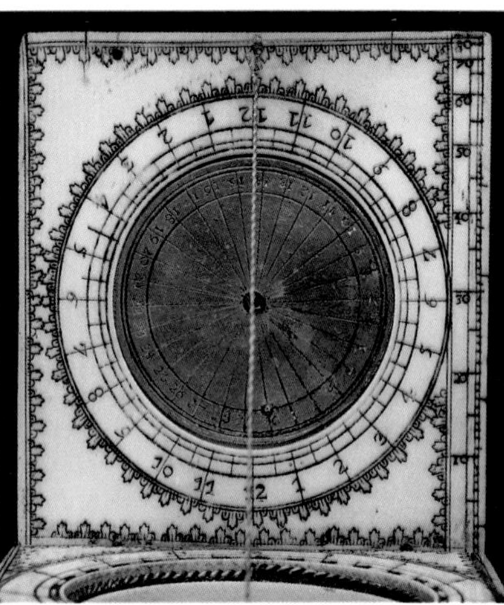

<table>
<tr><td>Ia</td><td>Ib</td></tr>
<tr><td>IIb</td><td>IIa</td></tr>
</table>

74 Rectangular Ivory Diptych

unsigned
mid to late 17th century
Dieppe, France
5.1 (w) x 5.7 (l) x 1.1 (h) cm
Formerly Harold Gillingham Collection
No. 90, then David P. Wheatland
Collection
Inventory No. 7502

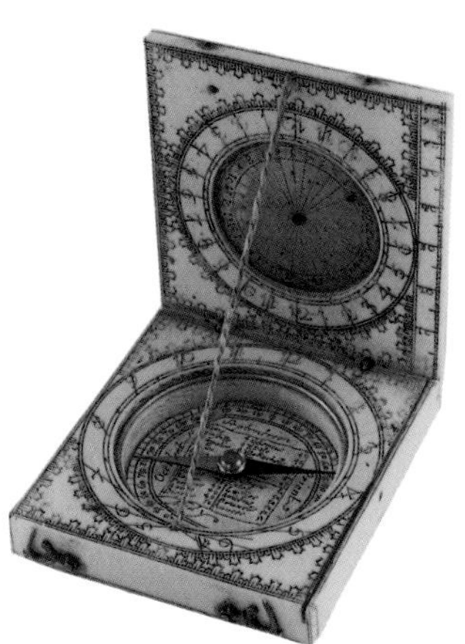

● Black coloring. Silver scythe clasps and hinges. Double-lined border on the outer edges of I and II. Hole in left side of II for pin gnomon (missing). Clasp also missing. Silver arm at right side of IIa to set upper plate to the proper angle for equinoctial dial. Compass needle is probably a replacement. Glass missing. Silver ring to hold compass glass intact.

Ia Double-lined border with typical Dieppe "pecked" ornamentation surrounds a circular equinoctial dial labeled 1–12 twice for use in spring and summer months. Central metal rivet with hole for pin gnomon.

Ib Double-lined border with typical Dieppe "pecked" ornamentation surrounds silver type 2 French lunar volvelle (badly worn) calibrated 1–30. Outer scale calibrated 1–12 twice, also to be used as scale for equinoctial dial during autumn and winter months. Latitude scale along right side from 0° to 80° to set angle for equinoctial dial.

IIa Horizontal dial with a single scale for approximate latitude 49°, labeled 5–12–7. Double-lined border with typical Dieppe "pecked" ornamentation. Printed paper latitude list of 24 cities, mostly French (42°–51° latitude), inside compass bowl. Directions labeled *Septentrion, Orient, Midy,* and *Occident.* Outer scale on paper insert calibrated 0°–90° four times, beginning with north and south.

IIb Plain, except for double-lined edging with Dieppe "pecked" ornamentation.

75 Rectangular Ivory Diptych

unsigned
mid to late 17th century
Dieppe, France
5.8 (w) x 6.6 (l) x 1.5 (h) cm
Formerly David P. Wheatland Collection
(purchased in France, 1952)
Inventory No. 7512

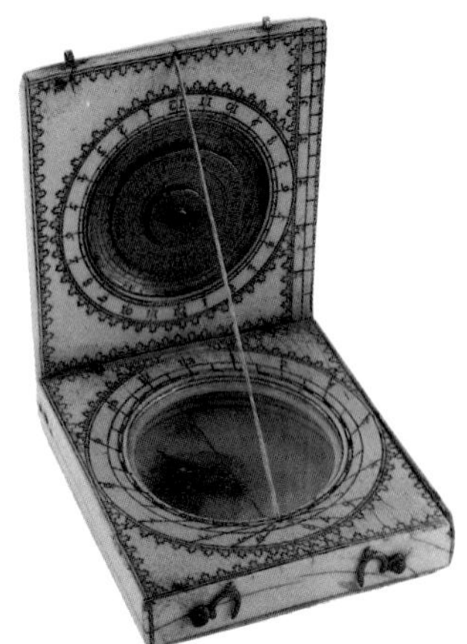

⚙ Black coloring. Brass scythe clasps (one with iron nail, one gilt eye) and hinges. Double-lined border on the outer edges of I and II. Hole in left side of II for pin gnomon (missing). Clasp also missing. Brass arm on right side of IIa to set angle for equinoctial dial. Hole in II, pieced together with wax. Compass needle and glass missing.

Ia Double-lined border with typical Dieppe "pecked" ornamentation surrounds a circular dial labeled 1–12 twice for use in spring and summer months. A brass indicator (later addition) folds up, possibly for use as a gnomon; setting the indicator to a specific hour moves the volvelle on tablet Ib.

Ib Double-lined border with typical Dieppe "pecked" ornamentation surrounds pewter type 1 French lunar volvelle of three discs (appear to be a crude replacement). Scales labeled 1–12 twice (outer, also to be used as scale for equinoctial dial during autumn and winter months), 1–12, JANV–DECE (middle), and 1–31 (inner). Latitude scale calibrated from 0° to 80° degrees along right side to set angle for equinoctial dial.

IIa Horizontal dial with a single scale for approximate latitude 49°, labeled 4–12–8. Double-lined border with typical Dieppe "pecked" ornamentation. Compass bowl empty.

IIb Plain, except for double-lined border along edge with typical Dieppe "pecked" ornamentation. Large crack and hole near center filled in with wax.

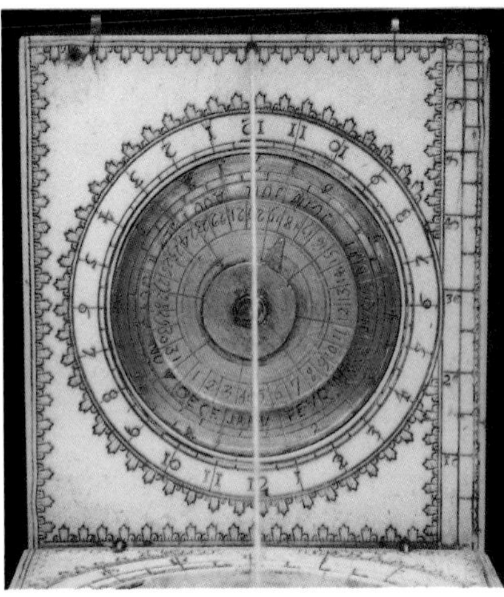

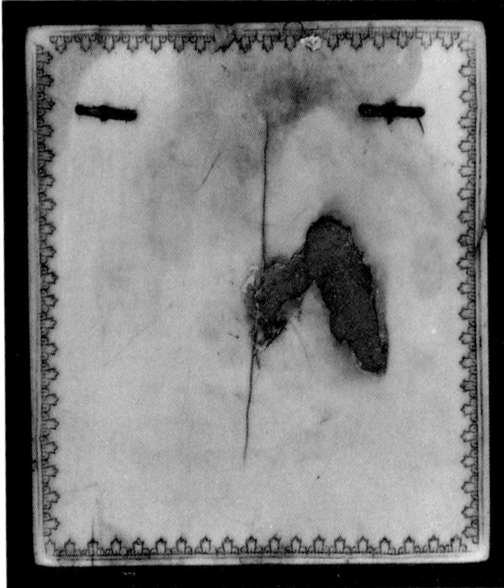

Ia | Ib
IIb | IIa

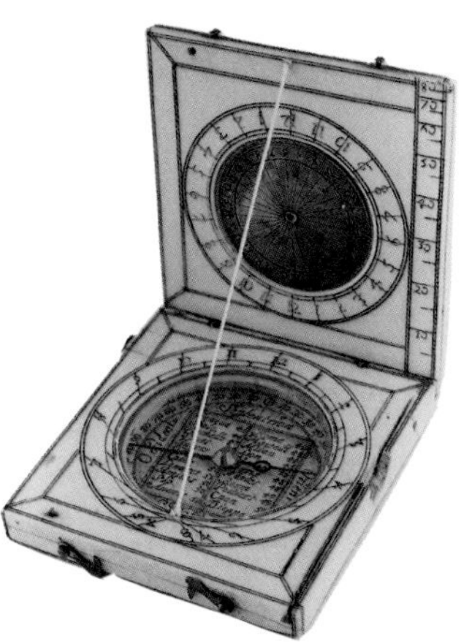

76 Rectangular Ivory Diptych

unsigned
mid to late 17th century
Dieppe, France
5.2 (w) x 5.7 (l) x 1.0 (h) cm
Formerly Drecker Collection No. 160,
then David P. Wheatland Collection
Inventory No. 7505

◉ Black coloring. Brass arm at right side of IIa to set angle for equinoctial dial. Two brass hook clasps on II. Compass needle and glass intact. Brass wire hinges. Blued-steel compass needle with gilded brass center. Cast brass clasp on left side of II to hold gnomon rod (intact). Double-lined border on outer edges of I and II. See figure 1, page 24.

Ia Double-lined frame with no ornamentation. Circular equinoctial dial labeled 1–12 twice, for use in spring and summer months; polar dial inside labeled 8–12–4. Central metal rivet with hole for pin gnomon.

Ib Double-lined frame with no ornamentation. Simple pewter type 2 French lunar volvelle with scales labeled 1–30 (inner) and 1–12 twice (outer, on ivory, also to be used as equinoctial dial scale in autumn and winter months). Latitude scale along right side from 0° to 80° to set angle for upper tablet.

IIa Horizontal dial with a single hour scale for approximate latitude 48.5°, labeled 4–12–8. Printed paper insert (painted yellow, red, and blue) in compass bowl with latitude table for 22 cities, mostly French (42°–51° latitude). Outer scale on paper insert calibrated 0°–90° four times beginning at north and south. Double-lined frame with no ornamentation.

IIb Double-lined frame with no further ornamentation. Number 4 written in red ink.

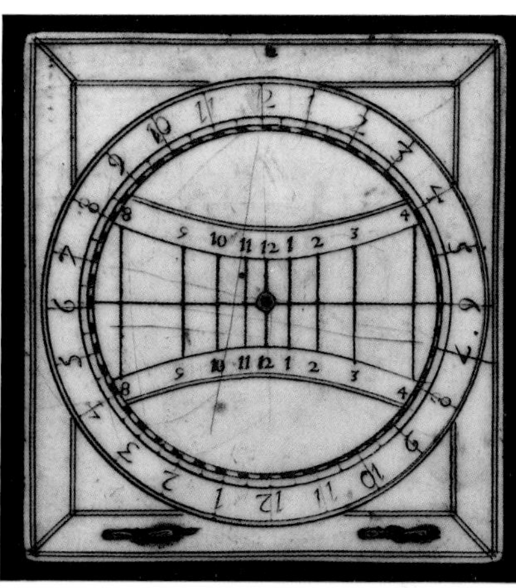

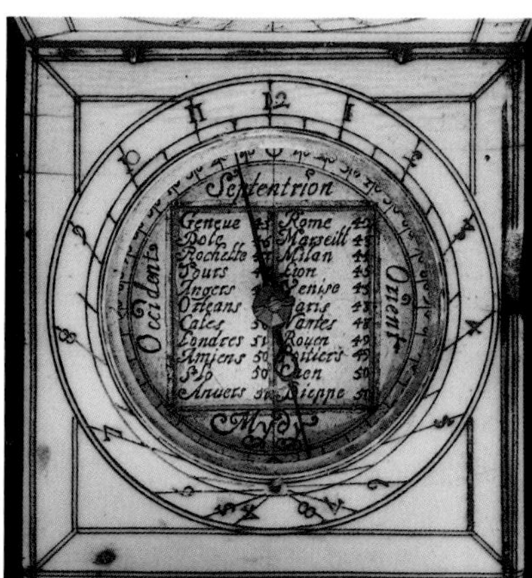

Ia | Ib
IIb | IIa

77 Miniature Ivory Diptych

unsigned
mid to late 17th century
Dieppe, France
4.2 (w) x 5.0 (l) x 1.1 (h) cm
Formerly Harold Gillingham
Collection No. 98, then David
P. Wheatland Collection
Inventory No. 7506

⚙ Black coloring. Brass scythe clasps and hinges. Double-lined border on the outer edges of I and II. Brass-clasped hole in left side of II for gnomon rod (missing). Brass arm at right side of IIa to set upper plate to the proper angle for equinoctial dial. Compass needle (replacement?), glass, and silver ring intact.

Ia Double- and single-lined border with poppy-flower ornamentation in corners surrounds a circular equinoctial dial labeled 1–12 twice (for use in spring and summer months) and central polar dial labeled 9–12–3. Small floral stamps between polar and equinoctial dials. Central metal rivet with hole for pin gnomon.

Ib Double- and single-lined border with poppy-flower ornamentation in corners surrounds pewter type 2 French lunar volvelle calibrated 1–30. Outer scale on ivory calibrated 1–12 twice, also to be used for equinoctial dial in autumn and winter months. Latitude scale along right side from 0° to 80° to set angle for equinoctial dial.

IIa Horizontal dial with a single scale for approximate latitude 48.5°, labeled 5–12–7. Double-lined border with poppy-flower ornamentation in corners. Printed paper insert in compass bowl has a latitude list of 18 cities, mostly French

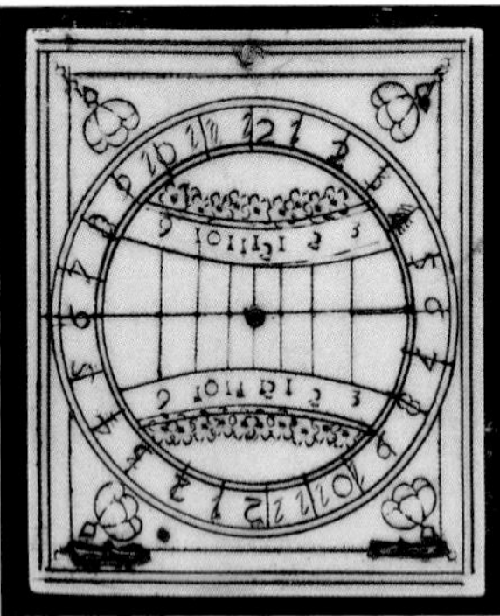

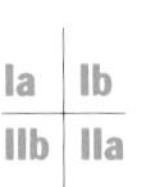

Ia	Ib
IIb	IIa

(16°–56° latitude). Directions labeled *Septentrion*, *Orient*, *Mydy*, and *Occident*. Outer scale on paper insert cut off.

IIb Double- and single-lined border with decorative pattern.

78 Miniature Ivory Diptych

unsigned
mid to late 17th century
Dieppe, France
4.2 (w) x 5.0 (l) x 1.3 (h) cm
Formerly Drecker Collection
No. 174, then David P.
Wheatland Collection
Inventory No. 7496

⊛ Black coloring. Brass scythe clasps and hinges. Double-lined border on the outer edges of I and II. Brass-clasped hole in left side of II for pin gnomon (missing). Silver arm (broken) at right side of IIa to set upper plate to the proper angle for equinoctial dial. Compass needle and glass missing.

Ia Double-lined frame with foliate ornamentation surrounds a circular equinoctial dial labeled 1–12 twice, with half-hours marked, for use in spring and summer months. Central metal rivet with hole for pin gnomon.

Ib Double-lined border with foliate ornamentation surrounds brass type 1 French lunar volvelle calibrated 1–12 twice (outer fixed disc) and 2–30 in two's (inner moveable disc). Scale for equinoctial dial beneath volvelle on ivory reads 6–12–6, to be used in autumn and winter months. Latitude scale along right side from 0° to 80° to set angle for equinoctial dial.

IIa Horizontal dial with a single scale for approximate latitude 49.5°, labeled 4–12–8. Double-lined border with decorative ornamentation. Compass bowl is empty, except for inscribed numeral 2. Modern paper insert.

IIb Plain, except for double-lined border. Ivory noticeably yellowed.

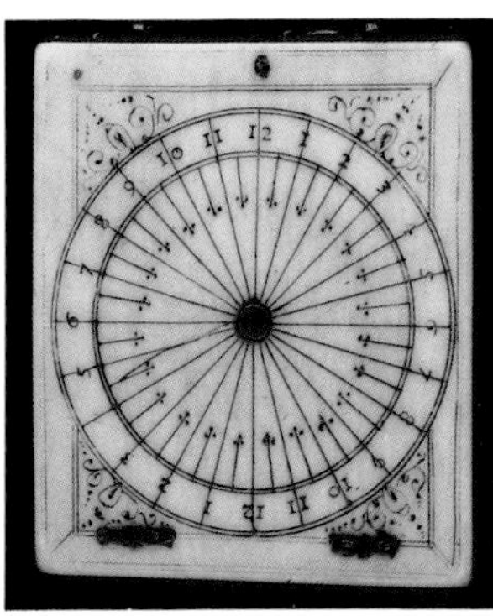

Ia | Ib
IIb | IIa

4

Flemish and Italian Diptych Sundials

Hanc itaq; regulam pone super centrum rotulæ, & fac vnum foramen rotundum, in quod mittatur clauus cum foramine rotundus, qui ambo constringat, ita vt index hac atque illac volui possit.

Vt patet in figura præsenti.

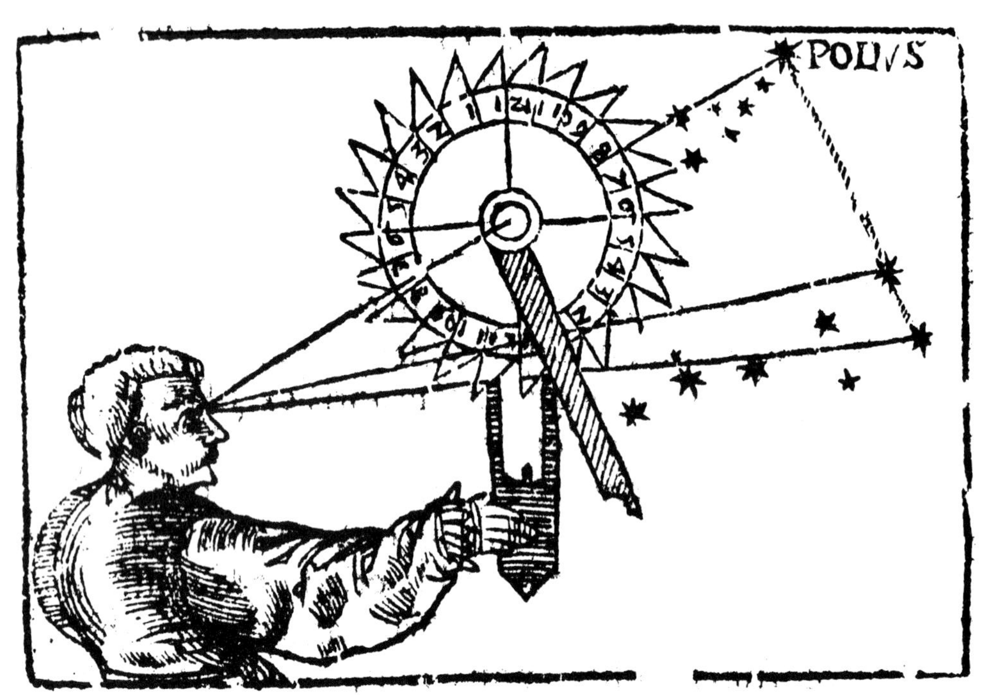

Usus huius Instrumenti.

I N stellifera noctis claritate subleua rotulam cum manubrio versus Septentrionem, & moue rotulam huc & illuc, donec radius visualis ab oculo tuo per centrum instrumenti ad stellam polarem transeat: similiter extra rotulam, idem radius ad extremas duas stellas maioris Vrsæ, seu rotas plaustri protendatur, & tam diu leua aut summitte indicem, donec eius linea fiduciæ super radialem lineam incidat. Tunc enim considera horam limbi & eius partem, quam regula abscindit, cum qua intra figuram præcedentem, hoc pacto: Pône indicem maioris vrsæ siue plaustri super horam inuentam, indice inuariato manente, situetur Solis sua fiduciæ linea super diem tuæ considerationis, & ostendit in inferiori parte limbi horam quæsitam, quâ inuenire oportebat.

Libri de Geographicis principijs
Finis.

Ivory Diptych Sundials in Flanders

by A. J. Turner

In Nuremberg the making of ivory diptych dials was a full-scale trade of which the products spread throughout Europe. In France, although the quantity of dials produced was too small to be an independent trade, dial-making nonetheless constituted a distinctive, specialized, part of ivory-working. In Flanders, by contrast, the making of ivory dials is almost invisible. That a few such dials were made is known by the fact that six have survived, but concerning them there is virtually no information.

Of the six known Flemish dials only one is signed and dated. This is a multiple dial signed "Factum per Victorem de Dayn Gandensem"[1] and dated 1549. It incorporates a 32-point wind rose, vertical and horizontal dials with a common string gnomon, a shadow square, a degree scale, and horary diagrams for equal, unequal, and Italian hours. In addition, the dial is rubricated with legends, biblical citations, and mottos. Both by its date and its style this dial separates itself from the other five known, none of which is signed, although all are dated between 1586 and 1599. Harvard's sole Flemish sundial, dated 1599, is one of these five (cat. no. 79),[2] and it is almost identical with the 1598 dial in Bielefeld.

The dials of this group are remarkably homogeneous in style and layout. All are multiple dials carrying on face Ia scales for a nocturnal with a brass conversion disc for solar and sidereal time and a hinged brass arm. Face Ib carries an equinoctial dial surrounding a lunar age scale, a lunar volvelle, and an engraved diagram showing the aspects of the planets. Along the right-hand edge there is a latitude scale for setting the leaf in the equinoctial plane against a brass strut hinged on the side of IIa and an attachment for the string gnomon used with the horizontal dial, marked in a rectangle with arched top around the inset compass in face IIa. Face IIb is entirely taken up with a latitude table, of which the Flemish forms of the names marked supply the most direct means of locating the dials geographically. A formal, symmetrical, rather stiff leaf and scroll decoration is also common to the dials.

Of the origin of these dials we know nothing. To consider their design as an adaptation of Nuremberg models is not unreasonable,[3] although the position of the lunar volvelle in face Ia might rather suggest a French influence, and the regular incorporation of a nocturnal seems to be an original feature. That they were manufactured in Antwerp as is sometimes suggested[4] is a conjecture that is supported by no evidence beyond the fact that Antwerp is usually the first place name in the latitude list (which is

1. "Made by Victor de Dayn of Ghent." The dial is now in a private collection. For de Dayn, publisher, printer, and engraver, see Bergmans (1890–1891).

2. The other four dials are dated and located as follows: 1586, Lewis Evans Collection, Museum of the History of Science, Oxford, M.9; 1595, British Museum, London, CAI/2414, Ward (1981), no. 89, p. 39; 1598, Huelsmann collection, Bielefeld, Inv. H-W 18a and b, see Syndram (1989), no. 17, pp. 85–86; and 1599, private collection.

3. The layout of face IIa recalls that of a dial dated 1598/9 and ascribed to Paulus Reinmann; Gouk (1988), p. 119, no. 1.

4. Syndram (1989), p. 86.

"

arranged neither alphabetically nor in ascending or descending order of the latitude values). That Antwerp by 1600 was probably the main center of Flemish instrument-making increases the probability that the dials were made in that city, as does a similarity between the *V* and the *R* of *VRSA* on the nocturnal arm with the *V* and *R* of *Vmbra Recta*, which are the only Roman capital letters engraved on some instruments by Michel Coignet, the leading Antwerp instrument-maker.[5] If the ivory dials were made by specialist craftsmen then it is perhaps not unlikely that the brass fittings were made and engraved in the workshop of a mathematical instrument-maker like Coignet. That the ivory dials may have been made by general instrument-makers who (like their Nuremberg and Paris counterparts) worked in a variety of materials is, however, suggested by the existence of a similar though rather more elaborately decorated dial, dated 1597, which is made of gilt brass.[6]

But on the basis of a minuscule sample it would be unwise to speculate further. That the five ivory dials are to be studied as a single group is clear. That they originated in the same town or region is highly probable. It is not unlikely that they were made in the same workshop or group of workshops. Beyond this at present we cannot go. Nonetheless, it is worth identifying this independent group of ivory dials which, however small, should be set alongside the more important French and Nuremberg manufactures. Eventually perhaps a broader study may be able to identify a comparable group or groups of Italian dials,[7] although it seems unlikely that the manufacture of either these, the Flemish dials, or those produced elsewhere in Europe ever developed into a full-scale trade like that of Nuremberg or of the French makers.

5. For example, on the surveyor's circle described by Chapiro, Meslin-Perrier, and Turner (1989), no. 91, p. 128.

6. Whipple Museum of the History of Science, Cambridge, Inv. Wh 1712; Bryden (1988), p. 63.

7. For some examples, see cat. nos. 80–82.

79 Rectangular Ivory Diptych

unsigned
dated 1599
Antwerp, Flanders
4.9 (w) x 6.9 (l) x 1.5 (h) cm
Formerly Harold Gillingham Collection
No. 86, then David P. Wheatland
Collection
Inventory No. 7527

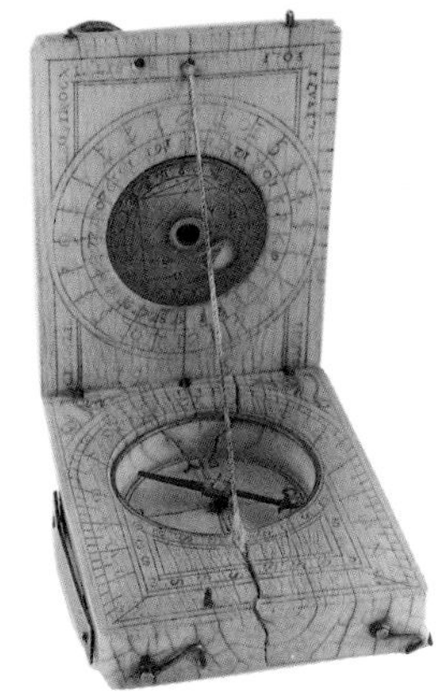

Faded black coloring. Tablet II is cracked lengthwise (glued back together?). Hole with long brass clasp in left side of II for gnomon rod (missing). Brass wire clasps on II. Compass needle intact, glass and brass ring missing.

Ia Brass nocturnal, labeled "VRSA MAIOR" (the Big Dipper) on the collapsible arm. Scales labeled "HORA VRSA MAIOR" and 4–12–8 (outer) and "HORA SOLIS" 8–12–4 (inner) on the volvelle. The inner hour scale, when positioned correctly, can be used as an equinoctial dial during spring and summer. Outer calendrical scale on ivory. Engraved scrollwork in corners.

Ib Engraved standard German lunar volvelle, with lunar phase indicated through hole in the rotating brass disc. Disc engraved with diagram for the aspects of the planets. Scales labeled 1–12 twice (inner and outer) and 1–29 (middle). Also meant for use as an equinoctial dial for the autumn and winter months. Labeled "CIRCVLIS AEQVINOCTIALIS" along left side, and "GRADI ELEVATO POLI" along the right. Latitude scale on right for 10°–80° to set the angle for upper tablet.

IIa Horizontal dial with a single scale for approximate latitude 51.5°, labeled 4–12–8, with lines marking the half- and quarter-hours. Ornamented with 20 letter-S figures. Engraved scrollwork in upper corner. Compass bowl has a paper insert, accented in red ink, with the cardinal directions. Magnetic declination 8° west of north.

IIb Latitude list of 14 European cities (42°–56° latitude); the first is Antwerp and the first four are in the Low Countries.

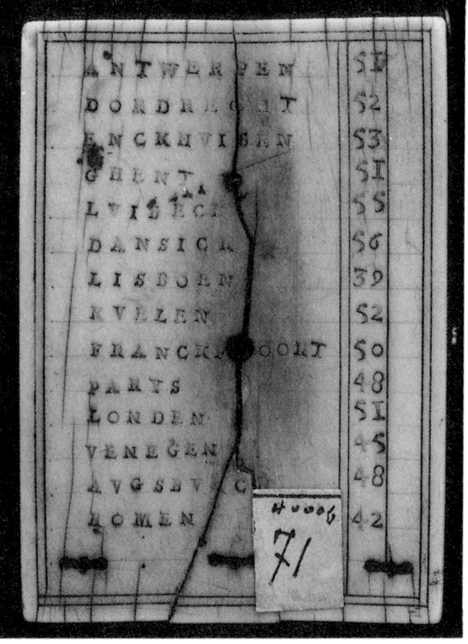

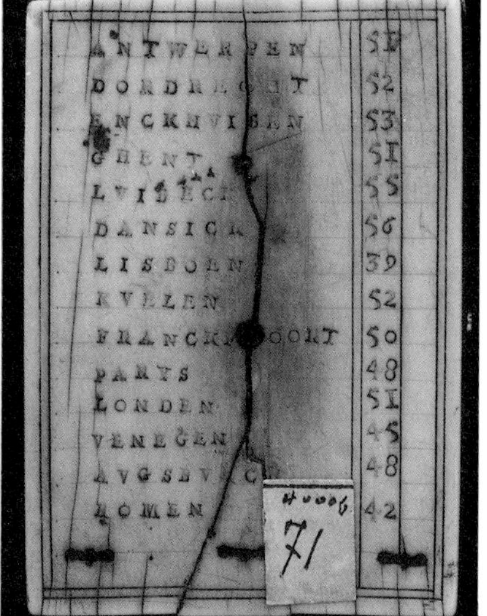

Modern paper label reads "4000f 71." Image inverted for legibility.

The origin of this diptych is unclear. Bielefeld's Kunstgewerbesammlung has an almost identical dial dated 1598, which their catalogue identifies as being from Antwerp. While Antwerp is indeed the first city listed on both dials' latitude tables, this in no way indicates the dials' city of origin. However, the linguistic forms of some of the place names make a Flemish origin certain.

Italian Diptych Sundials

There are three identifiably Italian dials in the Harvard Collection. Cat. no. 80 is almost certainly Florentine; its maker is unknown but it is signed with the initials D.G.B. It has a very unusual lunar volvelle made from pasteboard. Cat. no. 81, with its dual scaphe dials, is very similar to an ivory diptych at Chicago's Adler Planetarium; this dial is engraved with a cardinal's hat and the arms of the Colonna family. Cat. no. 82 bears the coat of arms of the Medici family.

The tiny pine tree below the fountain serves as a gnomon for the time on a scale calibrated in Italian hours. From Mario Bettini, *Recreationum Mathematicarum*, Bologna, 1659. By permission of the Houghton Library, Harvard University.

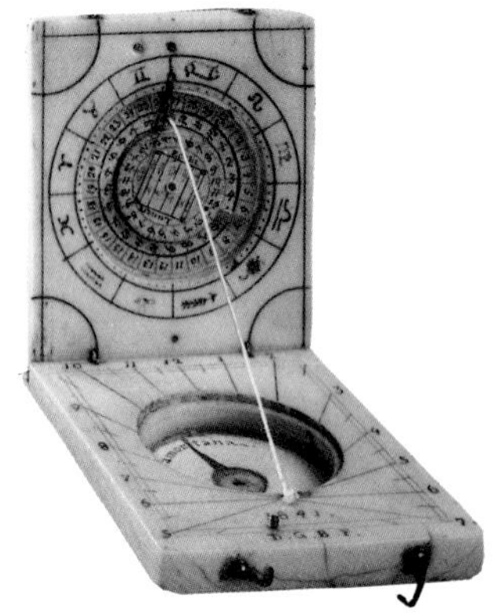

80 Rectangular Ivory Diptych

signed "D.G.B.F."
dated 1641
Florence, Italy
4.7 (w) x 5.8 (l) x 1.1 (h) cm
Formerly Drecker Collection No. 209,
then David P. Wheatland Collection
Inventory No. 7489

⬡ This small diptych is made from exceptionally smooth, creamy ivory. Red and black coloring. Brass hook clasps on I and II intact.

Ia Undecorated.

Ib An unusual lunar volvelle (functionally equivalent to the standard German lunar volvelle) made from pasteboard is recessed in the upper plate. Scales labeled 1–12 twice (inner and middle) and 1–29½ (outer). Zodiacal symbols are arranged concentrically around the volvelle. At the center of the volvelle is written "F. celit. Lunare" with a chart of the zodiacal symbols.

IIa Simple horizontal dial for approximate latitude 41.5°, labeled 5–12–7 in black numerals, with lines marking the half-hours. In compass bowl is a paper insert labeled "Tramontana" and "Mezzodi." Magnetic declination 20° west of north. Glass is a replacement. Inscribed "1641" and "D.G.B.F."

IIb Undecorated.

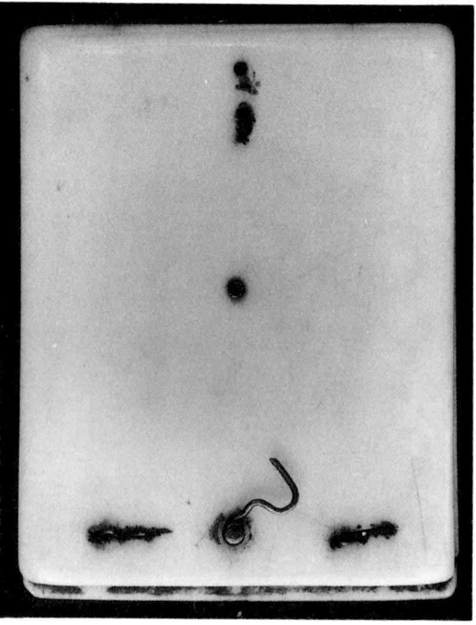

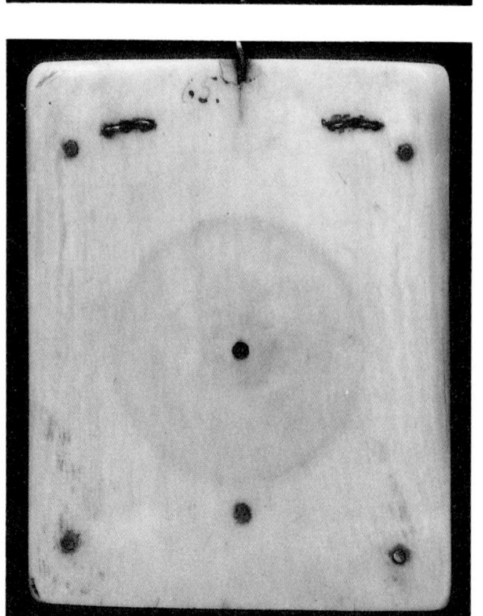

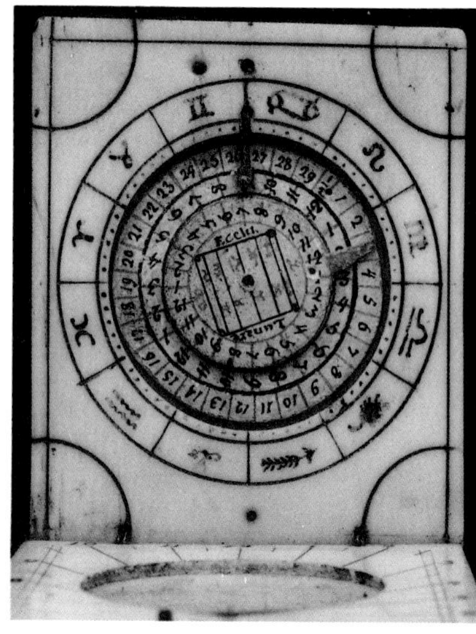

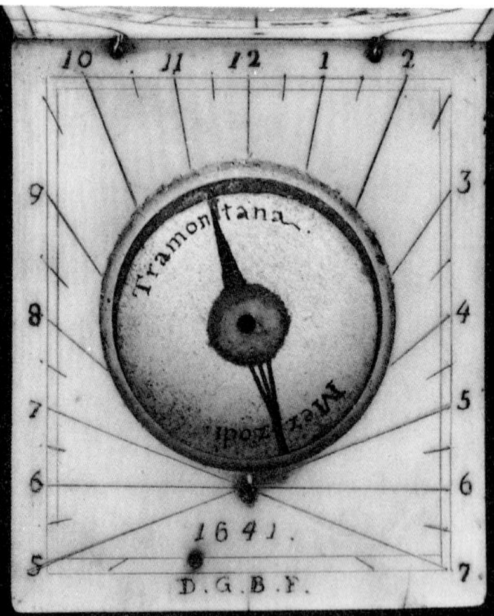

Ia | Ib
IIb | IIa

81 Ivory Diptych Book

unsigned
late 16th century
Italy
6.6 (w) x 7.8 (l) x 1.4 (h) cm.
Formerly Drecker Collection No. 208,
then David P. Wheatland Collection
Inventory No. 7488

This unusual diptych, in the form of a book, with gilt along the edges of the "pages," contains two scaphe dials for two different latitudes, recessed into very shallow hemispheres. Holes for plumb bob weights and guides for string suspensions on Ib and IIa. Brass bun feet on I, brass clasps on I and II missing, clasp catches and pin gnomons intact.

Ia Undecorated except for a single-line border.

Ib Scaphe dial labeled with Italian hour lines (9–23) and zodiacal symbols to represent the months. A compass bowl lacks both its pin and compass needle. Magnetic declination 20° west of north.

IIa Scaphe dial labeled with Italian hour lines (9–23) and another series of numbers (labeled 12–7), as well as zodiacal symbols. Again, the compass bowl lacks its pin and compass needle. Magnetic declination indicated for 6° east and 20° west of north.

IIb Undecorated except for a single-line border.

Compare with a similar dial, which is engraved with the arms of the Colonna family, in the Adler Planetarium (inv. no. W-490).

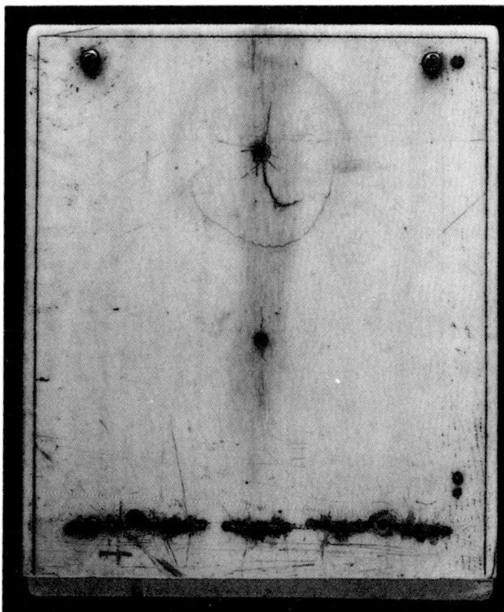

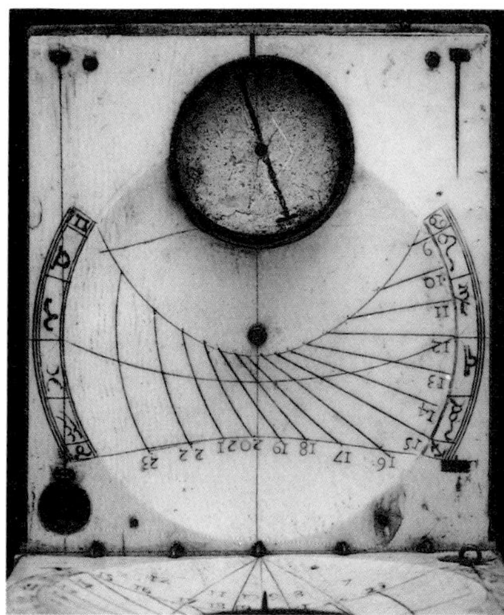

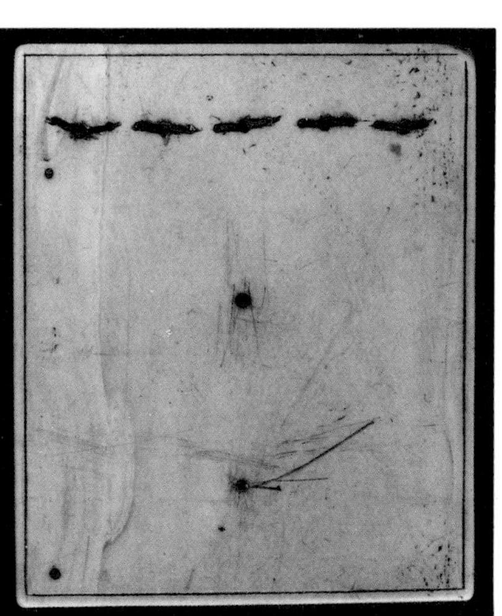

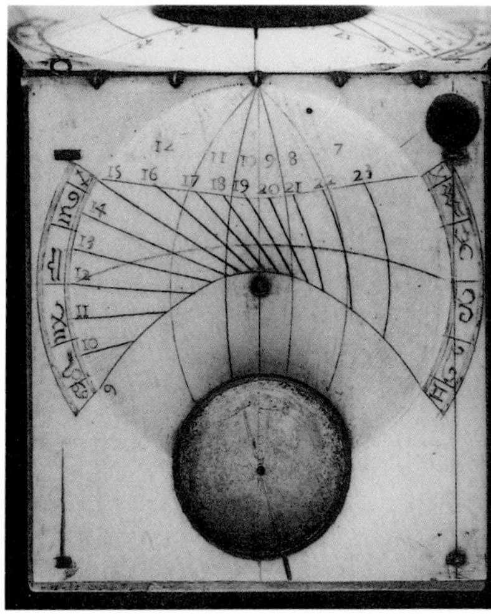

Ia	Ib
IIb	IIa

82 Rectangular Ivory Diptych

signed "GGF"
before 1569
Florence, Italy
4.3 (w) x 11.8 (l) x 1.7 (h) cm
Formerly Drecker Collection
No. 210, then David P.
Wheatland Collection
Inventory No. 7492

⬤ Black coloring. This ornate diptych is atypical in that it does not include a horizontal dial; it has only a single pin gnomon dial. Its primary function appears to be as a double compass, listing the directions of the prevailing winds, and the points of the compass. The crown over the Medicean coat-of-arms reveals that this dial was made before 1569 when the grand-duchy of Tuscany was formed. The signature "GGF" has not been identified. Silver hinge, clasp, and rings to hold glass (glass missing). Paper inserts in compass bowls.

Ia Decorated with the coat-of-arms of the Medici family of Florence. Silver clasp and hinge are elaborately ornamented. Image inverted for legibility.

Ib Decorated with an engraving of a man's face with leafy hair and a beard.

IIa Single pin gnomon dial with Italian hours labeled 12–23. Two compass bowls, both with hand-written paper inserts (possibly added later), the upper one labeled with 32 directions of the winds in Italian, and the lower one with 32 points of the compass in French. Engraved "GGF." Two different symbols for *Libra* on either side of the initials. Foliate decoration in the corners.

IIb Decorated with a simple geometric line engraving.

Ia | Ib
IIb | IIa

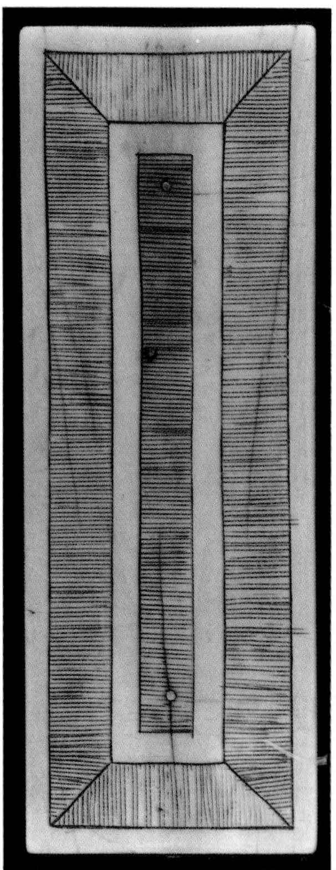
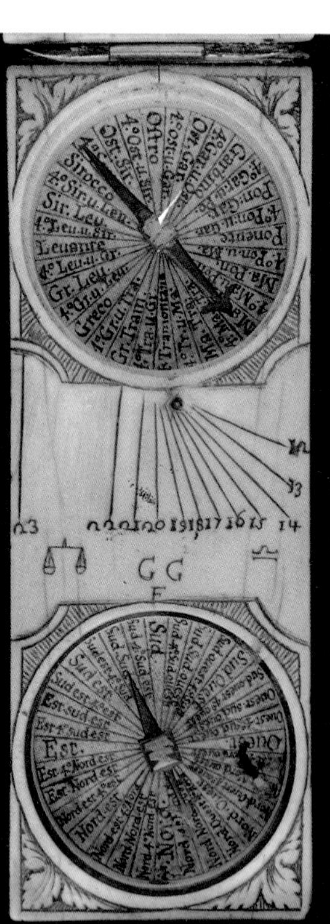

Appendices

PLAN PERSPECTIF DES CINQ HORLOGES

regulieres deffus nommées, lefquelles font iointes enfemble, & font faites fur les 60 degrez d'efleuation du pole.

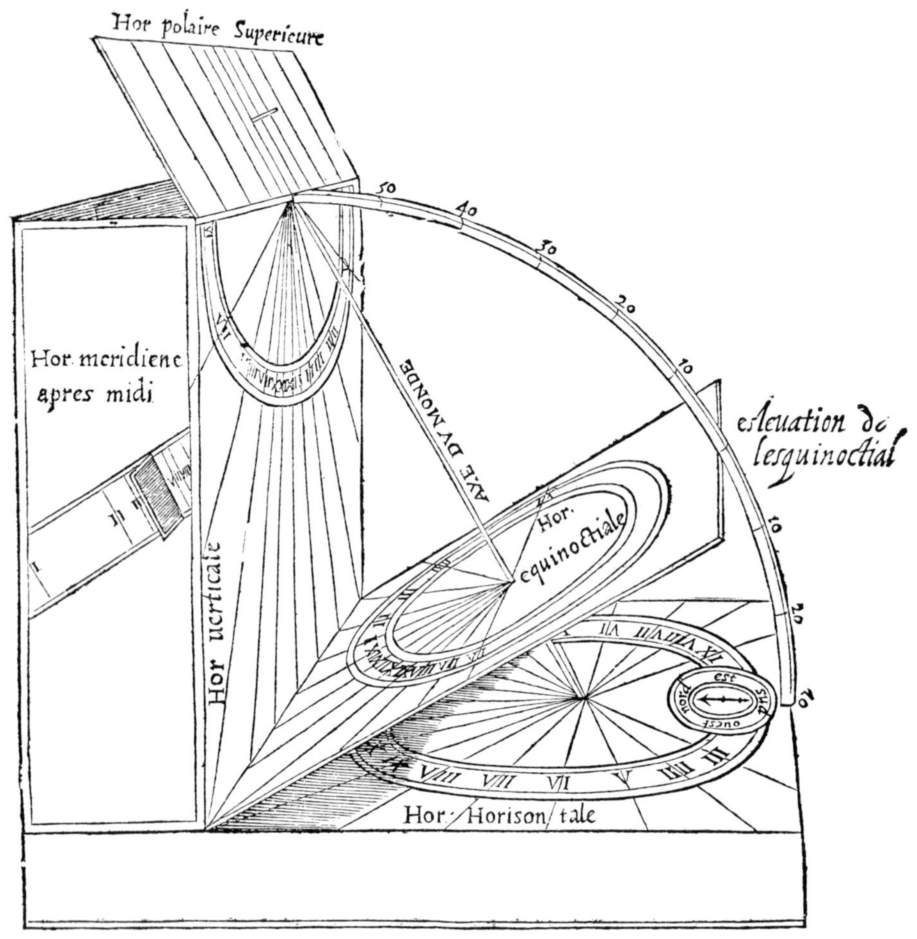

E plan cy deffus defigné donne vne forte intelligence des cinq horloges fufdites, où l'on void particulierement la difpofition en perfpectiue de chacune d'icelles, fauf la meridienne auant midy, la partie inferieure de l'efquinoxiale & la partie inferieure de la polaire : elles font toutes fur les 60 degrez d'efleuation du pole, qui font 30 degrez de l'efquinoxial.

L'on fait encores vne infinité d'horloges declinantes des fufdites, comme quand vne muraille verticale ne regarde pas iuftement le midy, ou quand vne autre n'eft pas paralelle au meridien, tellement qu'on regarde de combien elles declinent de degrez, comme fera enfeigné cy apres.

DEFINITION XXVIII.

IOVR NATVREL *eft le temps que le foleil demeure à faire fa circonference depuis vn midy iufques à l'autre.*

DEFINITION

Appendix I
Museums with Significant Collections
of Ivory Diptych Sundials

From Salomon de Caus, *La Pratique et demonstration des horologes solaires*, Paris, 1624. By permission of the Houghton Library, Harvard University.

Appendix II
The Dating of Sundials by Hans Ducher and Leonhart Miller

By using as references ivory diptychs that are *both* signed and dated, one can attempt to correlate the various signatures and abbreviations (as well as characteristic maker's marks) appearing on the instruments with the dates of their manufacture. The following is an attempt to establish such a correlation for dials signed Hans Ducher (or its variations) and Leonhart or Lienhart Miller. This comparison was significantly helped by access to Penelope Gouk's extensive set of photographs of Nuremberg ivory diptychs in European collections. Although this listing is by no means exhaustive, it does represent data from most of the larger public collections. (*Note:* The abbreviations of names of collections are identified in Appendix I.)

Hans Ducher

The signature "HANS DVCHER" appears to be the earliest; it was used from the 1560s up to at least 1580. The signature "hans ducher" appears in the late 1570s, while "HANS TVCHER" dominates the 1580s. Finally, the abbreviation "H D" was used in the following decade, and "H T" thereafter.

Date	Signature	Collection and Inventory Number
1567	HANS DVCHER	OXF G-202
1574	HANS DVCHER	SML 1952-235
1579	HANS DVCHER	GIP Nr. 1
1580	HANS DVCHER	NMM D.61/ 36-75C
1578	hans ducher	OXF G-203
1579	hans ducher	GNM WI 1874
1580	hans ducher	KSB H-W7
1582	HANS TVCHER	BM 67 7-5 26
1586	HANS TVCHER	BNM Phys 1
1588	HANS TVCHER	OXF G-204
1589	HANS TVCHER	OXF 57-84/77
1595	H D	CHSI 7574
1595	H D	OXF G-206
1595	H D	OXF G-207 (polyhedral)
1595	H D	TM 1201
1597	H D	Adler DPW-29 (triptych)
1600	H D	OXF 57-84/76
1614	H T	BNM 33/222

Leonhart Miller

Date	Signature	Collection and Inventory Number
1613	LIENHART MILLER	CHSI 7459
1613	LIENHART MILLER	CAM 1685
1613	LIENHART MILLER	OXF G-254
1616	LIENHART MILEL	CHSI 7565
1616	LIENHART MILER	Adler P. Tajan, 1979
1619	LIENHART MILER	NMM D.37/36-122C
1621	LIENHART MIELER	GNM WI 1032
1625	LINHART MIELER	Sotheby's, London, 5 May 1989
1627	LIENHART MILLER	Sotheby's, London, 5 May 1989
1627	LIENHART MILLER	OXF G-256
1628	LIENHART MNLLER	NMM D.43/36-128C
1629	LIENHART MILLER	CHSI 7560
1630	LEONHART MILLER	CHSI 7567
1634	LEONHART MILLER	GIP Nr. 5
1636	LEONHART MILLER	CHSI 7568
1636	LEONHART MILLER	KSB H-W9
1638	LEONHART MILLER	OXF G-258
1638	LEONHART MILLER	NMM D.239/36-207C
1638	LEONHART MILLER	Sotheby's, London, 5 May 1989
1639	LEONHART MILLER	OXF 57-84/85
1639	LEONHART MILLER	OXF 57-84/217
1640	LEONHART MILLER	CHSI 7562
1640	LEONHART MILLER	CAM 1684
1641	LEONHART MILLER	PRA č. 64396
1641	LEONHART MILLER	NMAH 326,979
1644	LEONHART MILER	CHSI 7563
1645	LEONHART MILER	NMAH 327,297
1645	LEONHARE MILER	KSB H-15
1646	LEONHART MILLER	CAM 184
1646	LEONHART MILLER	Adler W19
1647	LEONHART MILNER	MVW 176
1649	LEONHART MILRE	SML 1938-414
1649	LEONHART MILLER	CAM 1683

As the following list makes clear, there does appear to be a temporal distinction between dials signed "LIENHART MILLER" and those signed "LEONHART MILLER." But this in no way proves that there was more than one maker, since variations in spelling were commonplace at the time; for example, there are more versions of the family name, "Miller," than there are spellings of the maker's given name.

Glossary

altitude   The angular distance of a celestial or other object above the horizon.

altitude dial   A sundial in which the time is determined by the altitude or angular position of the sun above the horizon.

astronomical compendium   An instrument designed to serve a variety of astronomical functions; a composite of many smaller instruments placed together on a single, hand-held device.

autumnal equinox   The equinox (*q.v.*), occurring approximately 23 September, that marks the beginning of autumn in the northern hemisphere.

azimuth angle   The angular distance of a celestial or other object along the plane of the horizon, with respect to some reference direction (usually north).

Babylonian hours   A system of large hours (*q.v.*) that counts from 1 to 24 beginning at sunrise.

Bohemian hours   Italian hours (*q.v.*).

bun feet   Small, round or oblate balls of metal, usually brass, placed at the corners of some diptychs such that when the instrument is placed on a table or other surface, the outer face of the dial and the accompanying volvelles do not come into contact with the surface (to avoid scratching the ivory and metal parts).

cardinal points or **directions**   The four principal directions on a compass, mutually perpendicular, labeled north, east, south, and west.

celestial equator or **equinoctial**   The circle formed on the celestial sphere by the imaginary projection of the plane of the earth's equator into space.

celestial meridian   A circle formed by the intersection of the plane of a terrestrial meridian extended into space and the celestial sphere.

celestial poles   The two points on the celestial sphere defined by the imaginary extension of the earth's axis into space.

co-latitude   The angular distance representing the position of an object, measured from either pole (0°) to the equator (90°) along a meridian (*q.v.*), numerically equal to 90° minus the latitude.

common hours   Small hours (*q.v.*).

compass bowl   A cylindrical hollow in tablet IIa of a diptych containing a compass needle resting on a metal pin. Compass bowls are usually marked with the cardinal directions (*q.v.*) and other identifying marks.

compass needle   A magnetized iron or steel needle, with a metal (usually brass) central pivot, designed to indicate magnetic north (*q.v.*) when resting on a metal pin in the compass bowl.

declination   The angular distance of the sun or other celestial object north or south of the equator, measured along a great circle passing through the celestial poles (*q.v.*).

dial   Sundial; also occasionally used to refer to a volvelle (*q.v.*).

dial furniture   Features or component instruments on the surface(s) of a sundial.

dial plate   The surface of a sundial on which the shadow is cast and the hour lines are displayed.

diptych   An object made from two thin tablets, hinged at one end so that the tablets fold together.

diurnal period or **solar day**   The average length of time for the earth to rotate on its axis with respect to the sun (24 hours).

dominical or **Sunday letter**   The letter (A to G where A = 1 January) designating the date of the first Sunday in a given calendar year, used in the determination of the date of Easter; also for use with the perpetual calendar on Dieppe magnetic azimuth sundials.

ecliptic   The great circle (*q.v.*) traced out by the sun along its yearly path through the constellations. The ecliptic is inclined at an angle of about 23° 27' with respect to the plane of the earth's equator.

epact   A number in the 19-year metonic cycle that facilitates the intercalation between the solar and lunar calendars. More specifically, an epact aids in the calculation of the dates of moveable liturgical holidays (such as Easter and Pentecost, as defined by the lunar calendar) with reference to the fixed holidays (such as Christmas, as defined by the solar calendar).

equal hours   An hour system in which the diurnal period is divided into hours of equal length, typically into one group of 24 hours (large hours) or two groups of 12 hours (small hours).

equation of time   The difference (in minutes and seconds) between the local mean time (clock time) and the local apparent solar time (sundial time); this difference varies throughout the year from a maximum of +14.4 minutes in early February to a minimum of −16.4 minutes in early November.

equator   The terrestrial great circle (*q.v.*) the plane of which is perpendicular to the polar axis.

equinoctial or **equatorial dial**   A sundial that must be held with its dial plate perpendicular to the plane of the earth's axis and its gnomon parallel to the earth's axis (pointing to the celestial pole); the user orients it by inclining the dial northward (in the northern hemisphere) at an angle equal to the co-latitude. The hour lines proceed radially from the central gnomon and are spaced 15° apart. Equinoctial dials often have two dial plates, an upper one with hour lines for use in the spring and summer, and a lower one on the reverse for use in the autumn and winter.

equinox   Either of the two times each year (approximately falling on 21 March and 23 September) when the path of the sun crosses the plane of the equator; also, the dates when there are exactly twelve hours each of daylight and darkness over the entire globe.

foreign hours   Italian hours (*q.v.*).

French hours   Small hours (*q.v.*).

German hours   Small hours (*q.v.*).

gnomon   A rod or pointer; the shadow of its tip indicates the local apparent solar time.

golden number   The number of a particular calendar year in the 19-year metonic cycle (*q.v.*) used to determine the date of Easter.

great circle   The intersection of a plane and the surface of a sphere, when the plane passes through the center of the sphere; the shorter arc of the great circle between two locations on the surface of the globe is the shortest distance between these locations.

great hours   Large hours (*q.v.*).

Greek hours   Babylonian hours (*q.v.*).

Gregorian calendar   A revision of the Julian calendar (*q.v.*), introduced by Pope Gregory XIII in 1582 but not adopted in England and its American colonies until 1752.

hook clasp   A clasp made from a simple loop of metal.

horizontal dial   A sundial with a horizontal dial plate.

horsehead clasp   A stylized metal clasp whose shape somewhat resembles a horse's head.

hour lines or **hour scale**   The series of lines and accompanying numerical scales inscribed, written, or printed on the dial plate to indicate the time in a given hour system; the time is indicated by the position of the gnomon's or style's shadow with respect to the hour lines.

intercardinal points   The compass directions northeast, southeast, southwest, and northwest.

Italian hours   A system of large hours (*q.v.*) that counts from 1 to 24, beginning at sunset.

Jewish hours   Unequal hours (*q.v.*).

Julian calendar   A calendrical system, introduced in Rome in 46 B.C., of 365 days per year with an extra day each fourth year. The Julian calendar was in common use throughout Europe until the introduction of the Gregorian calendar (*q.v.*) in 1582, although it continued to be used in England and North America until 1752.

Kompassmacher or **Compassmacher**   The German term for a craftsman who manufactures sundials.

large hours   A system of equal hours (*q.v.*) in which the diurnal period is divided into 24 equal-length hours.

latitude   The angular distance of an object from the equator, measured northward or southward along a meridian from 0° at the equator to 90° at the poles.

liturgical calendar   The yearly cycle of moveable and fixed church holidays observed variously by the Roman Catholic and other Christian churches.

local apparent solar time   The time at the location of an observer as defined by the angular position of the sun above the horizon.

local mean time   The local apparent solar time, plus or minus a daily correction for the equation of time (*q.v.*).

longitude   The angle measured from the earth's axis between the prime meridian (*q.v.*) and the meridian (*q.v.*) of the observer, measured eastward or westward from the prime meridian, up to 180°.

lunar volvelle   A volvelle (*q.v.*) which, when set to the age or phase of the moon, allows one to convert the time at night (as determined by the shadow cast by the moon's light on a sundial) to the corresponding local apparent solar time.

magnetic azimuth dial   A sundial which, when oriented in the direction of the sun, indicates the time by the position of the compass needle along a calibrated hour scale.

magnetic declination   The angular distance between geographic and magnetic north. Sometimes erroneously referred to as "magnetic deviation."

magnetic north   The direction indicated by a compass needle as the magnetic north pole, which is slightly offset from the geographic north pole (the axis of the earth's rotation).

maker's mark   A characteristic mark scratched, engraved, or punched into a sundial by master craftsmen to indicate the workshop in which the sundial was manufactured.

meridian   A great circle (*q.v.*) through the geographic poles of the earth.

meridies   Latin for "south"; the word also refers to midday when the sun passes due south.

metonic cycle   A lunar calendar in which the relative phases of the moon return to specified dates within the calendar each 19 years.

moondial   A sundial with an adjustable outer ring set to the phase of the moon; the dial directly determines the local apparent solar time at night by the shadow cast by moonlight on the outer ring's scale.

nadir   The point on the celestial sphere vertically below the observer; the point diametrically opposite the zenith (*q.v.*).

nocturnal   A hand-held astronomical device which allows one to calculate the time at night by the relative positions of certain stars in the Big or Little Dippers with respect to the pole star, Polaris.

Nuremberg hours   An equal-hour system that starts counting with 1 at sunrise and again with 1 at sunset; the daylight hours correspond to Babylonian hours and are occasionally labeled as such on diptychs.

occidens   Latin for "west."

oriens   Latin for "east."

pewter   An alloy of tin with various amounts of lead, brass, and/or copper.

pin gnomon   Typically a thin metal rod, often pointed, which indicates the time on a sundial by the tip of its shadow.

planetary hours   An hour system, based on astrological principles, which assigns each hour of the day in succession to one of the seven "planets" known to antiquity.

polar dial   A sundial that must be held with its dial plate parallel to the earth's axis by inclining the dial plate southward (in the northern hemisphere) at an angle equal to the latitude; the hour lines (*q.v.*) are parallel to one another and are arranged symmetrically around the 12 o'clock line, from which the gnomon (*q.v.*) stands vertically above the dial plate.

pole star or **north star**   The star, named Polaris, that is located on the celestial sphere at approximately the north celestial pole (*q.v.*); its altitude is approximately equal to the observer's latitude.

quadrant   A divided quarter-circle with sights and a plumb-bob string for determining the angular elevation of celestial or terrestrial objects.

scaphe dial   A sundial in the shape of a bowl or a shallow, hollowed-out circular recess, with a pin gnomon to indicate the time on hour lines (*q.v.*) inscribed on the inner surface.

scythe clasp   A stylized metal clasp in the shape of a scythe.

septentrio   The Latin word referring to the seven stars of the constellations of Ursa Major (the Big Dipper) and occasionally in reference to Ursa Minor (the Little Dipper). Hence used on a compass to signify the direction north and the north wind.

small hours   A system of equal hours (*q.v.*) in which the diurnal period is divided into two groups of 12 equal-length hours, typically starting at midnight and noon.

solstice   Either of the two times each year (approximately falling on 21 June, the first day of summer, and 22 December, the first day of winter) when the declination (*q.v.*) of the sun reaches a maximum; also the dates of the longest and shortest periods of daylight in the year.

string gnomon   A string which, when adjusted to the proper latitude (*q.v.*) and held taut, indicates the time by its shadow. More accurately called a "string style."

style   A shadow-casting edge or surface that indicates the local apparent solar time.

sundial   An astronomical device typically used to indicate the local apparent solar time by the position of a shadow cast by a rod or gnomon (*q.v.*) or a shadow-casting edge or style (*q.v.*).

temporal or **temporary hours**   Unequal hours (*q.v.*).

tidal volvelle or **tide dial**   A simple form of a lunar volvelle (*q.v.*) which, when set to the high or low tide for a given port on the full or new moon, allows one to determine the times of the subsequent tides throughout the month.

unequal hours   An hour system in which the day and night are divided into twelve hours each; the lengths of the day and night hours differ, except on the vernal and autumnal equinoxes.

vernal equinox   The equinox (*q.v.*), occurring approximately 21 March, that marks the beginning of spring in the northern hemisphere.

vertical dial   A sundial with its dial plate perpendicular to the ground.

volvelle   Round disc or discs, usually made out of metal, that can be rotated; on a diptych, the volvelle is typically inscribed with several numerical scales to aid in mathematical or astronomical calculations or conversions.

Welsch hours   Italian hours (*q.v.*).

wind rose   An accessory found on the outer face of many ivory diptychs indicating four or more directions, often laid out in a stylized star pattern. If the user orients the wind rose properly, by sighting the compass needle through the compass viewing-hole and aligning the dial accordingly, a wind vane (*q.v.*) will indicate the prevailing wind direction on the wind rose. Also called a "compass rose" or "mariner's compass."

wind vane   A moveable metal flag that spins freely on a small pin; when the vane is inserted into the central hole of a wind rose, the wind will orient the vane so that it indicates the direction of the prevailing wind.

zenith   The point on the celestial sphere vertically overhead the observer.

zenith distance   The angular distance of a celestial or other object from the zenith.

zodiac   A circular band on the celestial sphere extending 8° on either side of the ecliptic (*q.v.*). The zodiac is divided into 12 regions of 30° each. Each region bears the name and symbol (sign) of the constellation it contains.

zodiacal symbols   The symbolic abbreviations used to identify each of the 12 signs of the zodiac. The symbols are associated with mythological figures and creatures.

Bibliography

Anthiaume, A. 1920. *Evolution et enseignement de la science nautique en France et principalement chez les Normands.* 2 vols. Paris.

Baron, H. 1937. "Religion and Politics in the German Imperial Cities during the Reformation." *English Historical Review,* vol. 52, pp. 405–427, 614–633.

Bergmans, Paul. 1890–1981. Article on "de Dayn." In *Biographie nationale Belge,* vol. 11, cols. 601–602.

Bobinger, M. 1966. *Alt-Augsburger Kompassmacher.* Augsburg.

Bovill, E. E. 1969. *The Golden Trade of the Moors.* 2d ed. London.

Bryden, David J. 1988. *The Whipple Museum of the History of Science, Cambridge. Catalogue 6: Sundials and Related Instruments.* Cambridge.

Chandler, B., and **C. Vincent**. 1969. "Three Nürnberg Compassmachers: Hans Troschel the Elder, Hans Troschel the Younger, and David Beringer." *Metropolitan Museum Journal,* vol. 2, pp. 211–216.

Chapiro, Adolphe, Chantal Meslin-Perrier, and **Anthony Turner**. 1989. *Musée National de la Renaissance, Château d'Ecouen. Catalogue d'Horlogerie et des instruments de précision du XVIe au milieu du XVIIe siècle.* Paris.

Christie's, South Kensington. 14 April 1988. *Time Measuring Instruments from the Time Museum.* Lot 65.

de Beer, E. S., ed. 1955. *The Diary of John Evelyn.* 6 vols. Oxford.

Dutka, J. 1988. "On the Gregorian Revision of the Julian Calendar." *Mathematical Intelligencer,* vol. 10, no. 1, pp. 56–64.

Endres, R. 1970. "Zur Einwohnerzahl und Bevölkerungsstruktur Nürnbergs im 15./16. Jahrhundert." *Mitteilungen des Vereins für Geschichte der Stadt Nürnberg,* vol. 57, pp. 242–271.

Franklin, Alfred. 1906. *Dictionnaire historique des arts, métiers et professions exercés dans Paris depuis le treiziéme siècle.* Paris and Leipzig.

Garin, Eugenio. 1983. *Astrology in the Renaissance.* London.

Gouk, P. M. 1988. *The Ivory Sundials of Nuremberg, 1500–1700.* Cambridge.

Grandjean, Gilles. 1990. *Gloria à Rouen: Une famille d'horlogers au XVIIIe siècle.* Rouen.

Hartmann, J. 1919. *Die astronomischen Instrumente des Kardinals Nikolaus Cusanus.* Berlin.

Irmscher, G. 1985. "Der Nürnberger Ornamentstich im 16. u. 17. Jahrhundert." In *Wenzell Jamnitzer und die Nürnberger Goldschmiedekunst 1500–1700: Eine Ausstellung im Germanischen National-museum Nürnberg vom 28. Juni– 15. September 1985,* pp. 141–150. Munich.

Janin, L., and **D. A. King**. 1977. "Ibn al-Shāṭ'ir's Sanduq al-Yawaqut: An Astronomical Compendium." *Journal for the History of Arabic Science,* vol. 1, pp. 187–256.

Jegel, A. 1965. *Alt-Nürnberger Handwerksrecht und seine Beziehungen zu andern.* Neustadt an der Aisch.

Jones, Charles W., ed. 1943. *Bedae opera de temporibus.* Cambridge, Mass.

Kellenbenz, H. 1974. *Das Meder'sche Handelsbuch und die Welser'schen Nachträge.* Wiesbaden.

Le Corbeiller. 1914. "Les quaranniers Dieppois en 1662." *Bulletin Trimestriel des Amis du vieux Dieppe,* vol. 7, p. 64.

Le Monnier, Pierre-Charles. 1771. "Variations de l'aimant à Paris." *Histoire et Mémoires de l'Académie Royale des Sciences,* pp. 459–460.

L'Espinasse, René de. 1886–1897. *Les Métiers et corporations de la ville de Paris.* 3 vols. Paris.

L'Espinasse, René de, and **Francois Bonnardot**. 1879. *Le Livre des métiers d'Etienne Boileau.* Paris.

Lockner, H. P. 1981. *Die Merkzeichen der Nürnberger Rotschmiede.* Munich.

Maddison, Francis, and **Anthony Turner**. "Sun-Time, Moon-Time and the Compass: A Study of the Magnetic Azimuth Dial" (in preparation).

Malin, S. R. C., and **Sir Edward Bullard**. 1981. "The Direction of the Earth's Magnetic Field at London, 1570–1975." *Philosophical Transactions of the Royal Society of London. A. Mathematical and Physical Sciences*, vol. 299, pp. 357–423.

La Mesure du Temps dans les Collections Belges. 1984. Brussels.

Michel, Henri. 1939. *Introduction à l'étude d'une collection d'instruments ancien de mathématiques.* Antwerp.

Michel, Henri. 1953. *Les Cadrans Solaires de Max Elskamp, Musée de la Vie Wallonne.* Liège.

Milet, Ambroise. 1904. *Anciennes Industries scientifiques et artistiques Dieppoises.* Dieppe.

Milet, Ambroise. 1906. *Ivoires et ivoiriers de Dieppe, étude historique.* Paris.

Murdoch, Tessa. 1984. "Some Huguenot Craftsmen from Dieppe in London." *Seventeenth Century French Studies*, vol. 6, pp. 60–74.

Parry, J. H. 1963. *The Age of Reconnaissance, 1450–1650.* London.

Pilz, K. 1954. *Das Handwerk in Nürnberg und in Mittelfranken.* Nuremberg.

Ripa, Cesare. 1970. *Baroque and Rococo Pictorial Imagery*, trans. E. A. Maser. New York.

Schnelbögl, F. 1966. "Life and Work of the Nuremberg Cartographer Erhard Etzlaub († 1532)." *Imago Mundi*, vol. 20, pp. 11–26.

Schuhmann, M. R. 14 and 15 November 1935. *Objets d'art et d'ameublement, faiences et porcelaines anciennes . . . instruments de mathématiques composant la collection de M. Robert Schuhmann.* Lot 123. Paris.

Seebass, G. 1972. "The Reformation in Nürnberg." In *The Social History of the Reformation*, ed. L. P. Buck and J. W. Zophy, pp. 17–40. Columbus, Ohio.

Siraisi, Nancy. 1990. *Medieval and Early Renaissance Medicine: An Introduction to Knowledge and Practice.* Chicago.

Stockbauer, J. 1879. *Nürnbergisches Handwerksrecht des XVI Jahrhunderts.* Nuremberg.

Strauss, G. 1966. *Nuremberg in the Sixteenth Century.* New York and London.

Syndram, Dirk. 1989. *Wissenschaftliche Instrumente und Sonnenuhren.* Munich.

Tardy. 1972. *Dictionnaire des horlogers francais.* Paris.

Trombetta, Jean Pierre. 1987. "Excavation of the Louvre." *Atlas: In-Flight Magazine of Air France*, January, p. 139.

Turner, A. J. 1990. "The Origins of Modern Time." In *Time*, ed. A. J. Turner, pp. 18–21. The Hague.

Vincent, Claire, and **Bruce Chandler**. 1969. "Night-time and Easter Time: The Rotations of the Sun, the Moon, and the Little Bear in Renaissance Time Reckoning." *Metropolitan Museum of Art Bulletin*, vol. 2, pp. 372–384.

Wagner, H. 1901. "P. Apians Bestimmung der magnetischen Missweisung von 1532 und die Nürnberger Compassmacher." *Nachrichten der Gesellschaft der Wissenschaften zu Göttingen, I, Philologisch-Historische Klasse*, pp. 179–182.

Ward, F. A. B. 1981. *A Catalogue of European Scientific Instruments in the Department of Medieval and Later Antiquities of the British Museum.* London.

Wynter, Harriet. 1974. *Arts & Sciences: A Catalogue of Scientific Instruments*, vol. 2, no. 3. London.

Wynter, Harriet, and **Anthony Turner**. 1975. *Scientific Instruments.* London.

Zinner, E. 1956; reprint, 1979. *Deutsche und Niederländische Astronomische Instrumente des 11.–18. Jahrhunderts.* Munich.

Zinner, E. 1990. *Regiomontanus: His Life and Work*, trans. E. Brown. Amsterdam.

Inventory	Cat. No.	Page
7385	71	134
7386	70	133
7387	69	132
7388	65	128
7389	64	127
7390	68	131
7458	13	62
7459	16	67
7488	81	151
7489	80	150
7492	82	152
7493	47	106
7495	51	110
7496	78	141
7497	54	113
7498	55	114
7499	52	111
7500	53	112
7502	74	137
7503	72	135
7504	49	108
7505	76	139
7506	77	140
7507	62	125
7508	63	126
7510	61	124
7512	75	138
7513	58	121
7514	59	122
7516	73	136
7517	67	130
7518	66	129
7519	60	123
7521	40	92
7522	48	107
7524	35	88
7525	34	87
7526	36	89
7527	79	147
7532	15	64
7533	10	58
7534	12	60
7535	14	63
7537	11	59
7539	27	80
7540	33	86
7541	32	85
7542	26	79
7543	39	91
7544	41	92
7545	43	94
7546	37	90
7547	42	93
7548	45	95
7550	31	84

Inventory	Cat. No.	Page
7552	28	81
7553	29	82
7554	30	83
7559	46	96
7560	18	69
7561	25	76
7562	21	72
7563	22	73
7565	17	68
7567	19	70
7568	20	71
7570	23	74
7571	24	75
7573	8	55
7574	2	49
7575	3	50
7576	5	52
7577	4	51
7579	7	54
7581	6	53
7800	56	118
7830	1	48
7841	50	109
7887	44	94
7894	38	90
7897	57	120
7899	9	56

Index

HIBERNIA
OCEANUS GERMANICUS
ANGLIA
CANTABRICUS OCEANUS
HISPANIA
Mare Gallicum
Mare Ibericum
MARE
BARBARIA
Brazil
Tellin
Wisha
L. Spay
Iberlyn
Lista
TIA
Glasco
Edenburgh
Corswel
Barwyck
Kilton castel
Hoghmore
Carlnson
fort
Clare
Dublin
Wexford
Kensal
Dingle
Knos head
I. Man
Anglesey
Newrin
Denbig
Chester
Vorck
Hull
Lancaster
Ambridge
S. Davils
Sorlinges
S. Eurien
Notingam
Lincolne
Landof
Norwiche
Cambridge
Canterbury
Plymouth Bristol
Poole
Leewarden
Texel
Altmaar
Amsterdam
Hollandia
Rotterdam
Zelandia
Groningen
Zwol
Dartmout
Hersant
Diepe
Cales
Gent
Antwerpen
Bruxel
Pen
Luyck
Nam
Tur
West
Brest
Morlus
Britai
Dole
Blavet
Aure
ehe
Casn
Roan
Doay
Mons
Lutsenburg
Trier
Limburg
Coblints
Lavalnes
Norman
Noyen
Re: Ines
Vannes
gne
Nantes
Angiers
Chartres
Paris
France
Reyns
Verdun
Mets
Lorai
Worns
Loire
Poictou
Blois
Sens
Chalons
Champaigne
Straesburg
Talmont
Orlean
Troyes
La
Tubr
Poictiers
Rochfort
Langres
Els
Rochelle
Bourges
Nivernois
Bourbon
Nevers
Dole
Basel
Xanton
gne
Limoges
Bourgongne
Marennes
Angoulesme
Chalon
HEL
Bordeaux
Auvergne
Muscon
Lucern
Guienne
Quercy
Clermont
VE
Bayonne
Velay
Geneve
Losonna
Gual
Cahors
Lion
Augusta
TIA
Cogne
Mauban
Legur
Grenobl
Olaron
Tholouse
Valenee
Viviers
Langue
Mompeil
Aviñon
Pavia
doc
Agde
Arles
Turin
Narbo
Marseille
Savona
Provence
Tollon
Rosas
C. Dragonis
I. de Eres
Genava
C. d. Palos
Magorea
Balearides inl.
Yvica
Minorca I
I. Formentera
I. Cabrera
S. Petro
Tabarchia
Tunis
Zizana
C. Ortigal
Bello Isle
C. Finis terrae cum Promont.
Riba deo
C. d. Pinas
Villa viciosa
I. de Rez
I. d. Oleron
Santillana
S. Sebastian
I. de Baiona Minus fl.
Lugo
Galicia
Leon
Bragança
Bilbao
Palencia
Lagroño
Navarra
Durius flu.
Camora
Miranda
Burgos
Pampelona
Pori
Came
Valladolid
Jaca
Guarda
Cam
Castilia
Osuna
Combre
Salamanca
Avila
Gual
Lisbona
Guarda
Madrid
Siguença
Arragonia
Eboni
Placentia
Alcantara
Doroca
Saragoça
Tagus
Toledo
Cuenca
Pina
Lerida
Barcelona
Merida
Guadiana
Celta
Valen
Tortosa
Estremadura
Indivar
Ciudadreal
Carul
Peniscola
Cordua
Beca
Alacria
Sevilla
Boetis flu.
Sierra de
Valentia
Andaluzia
Granada
Alic
C. d. Gates
Guadalquivir R.
S. Lucar
Malaga
Baca
Cartagena
Calis I.
Ronda
Guad.
Almeria
ho de Gibraltar Gaditanum
Gibraltar
Salabreña
Yvica
Tanger
Alcazer
Albonan
Ruche
Farga
Tetuan
Veles
Melilla
Fez
Hone
Oran
Mostaban
Tenes
Alger
Bogia
Bona
Zerzeli